NOT QUITE RANDOM THOUGHTS

Howard S. Halpern, CCABW, is a retired applied physicist, engineer, sometimes chemist, whose career on behalf of the United States' national defense spanned 43 years.

NOT QUITE RANDOM THOUGHTS

Insights & Needles
from Haystacks

Howard S. Halpern

Longmeadow, MA

Copyright © 2017 by Howard S. Halpern
All rights reserved.
ISBN-13: 978-1545054697

Cover icons © Fotolia.com

Longmeadow, MA

Printed in the United States of America

This book is dedicated to my wife. From the moment I first laid eyes on her I was charmed. Her radiant beauty, joyfulness, and love of life continue to inspire me. My best decision in life was to propose to her. She has enabled me to be successful. She is a genius at relationships with people. In partnership, we raised a loving and capable family. Without her support and love this book never would have been written.

It is also dedicated to my family and future descendants who may disagree with me as I, in part, disagree with some of my ancestors whom I am sure would disagree with me.

CONTENTS

Preface	9
Introduction	13
1. Memory – the Importance of Forgetting	15
2. Aging – Your Health	21
3. The Creation of Time	39
4. The Nature of God – The God of Nature	49
5. UFOs	61
6. The Air We Breathe	73
7. Climate Squeal	77
8. Slowing Earth Rotation	87
9. Some Quirks of History	89
10. Evolution	129
11. A Little History of Nuclear Weapons	151
12. Principles of Mismanagement	165
13. Predictions	171
Afterword	173
About the Author	175

PREFACE

I am a retired applied physicist/engineer, sometimes chemist, who worked 43 years (1947–1990) on varied programs for the United States national defense, including delivery of nuclear weapons and defense against them. Just after World War II ended, I was a teaching assistant in chemistry and I had met the requirements for graduation. The following school year I was a teaching assistant in physics elsewhere. Then I went to work for a living. I have been a lifelong self-learner, mentored myself in science and non-science-based subjects.

My own work includes four patents. Many of my other projects were not patented due to the expense of patenting. Some of these projects were considered by the federal government as emergency measures which had to be done immediately. One such project was the first antenna used on the DEW Line which went from Alaska to Greenland as an early warning system in case the Russians would send bombers with nuclear warheads to attack the United States using this shorter route. Since the government felt this was urgent business, it violated its own laws because it could not wait for procurement of funds. Bell Telephone supervised this work to speed up the process. I worked for AMF which worked under contract for Bell. My job included testing for consistency in the production which was done in Connecticut and Brooklyn. The antenna was the size of a house.

Although most of my work involved the military, some of my work at AMF involved its bowling machines. There was a need for sensors to answer the question of pin location so machines could accurately assess how to pick up fallen pins without knocking others down. AMF is still a name commonly seen in bowling alleys.

I went to Key West to learn about naval operations, which led to my writing a classified report about the vulnerability of anti-submarine operations.

In Colorado, I worked with the air force on North American Radar Defense (NORAD). An invention of mine was 3D radar which was produced by United Technologies/Aircraft. This radar enabled planes to fly more safely at lower levels to avoid detection. It showed the terrain in three dimensions so pilots could avoid colliding with treetops, mountains, and other obstacles. This radar was used in the air force's planes used during the Vietnam War and the first Gulf War.

My work took me to Washington, DC in order to understand firsthand military needs and advise the government. Meetings at the Pentagon were classified, and so was a great deal of my work. There still exist many projects which require the world's attention for the security of society.

Throughout my life, I have written articles, some have been accepted for publication, some have not—this book includes some of my theories that were never published. It is my hope that future generations of researchers will be intrigued by my theories and will try to advance them. For ease of reading, there are no equations or footnotes in this book. It is written for people with open minds. It is meant to challenge mythology that is being taught as fact.

I cover many subjects: memory, aging and your health, the creation of time, the mass of the vacuum, is there an afterlife??, UFOs, the biochemistries of space aliens who evolved anywhere in the universe, and other quirks of history, evolution and brilliant design versus intelligent design, the male breast, Einstein, nuclear weapons, and free will. I give predictions and discuss principles of mismanagement.

These greatly varied thoughts are related by a way of thinking—like placing pieces in a huge jigsaw puzzle, relating thoughts. Sometimes like with a jigsaw puzzle turned upside down, starting in the corners with the blank pieces representing ignorance...

An objective of this book is to encourage the reader to think for him/herself. You are obviously free to disagree with what I have written here. I reserve the right to change my mind as long as I am alive and functioning. An old joke is, "It does you no good to agree with me; I have already changed my mind."

INTRODUCTION

This book is written and edited largely from memory. I am seeking to avoid telling you more than you would want to know, to avoid burying ideas under details, like hiding needles in a haystack. But I do digress, to offer examples to help you understand.

I would have hoped that in this day and age, I would be using a different form of the alphabet for writing this book. I feel that the present day alphabet is outdated. British author, George Bernard Shaw, wrote the book *Pygmalion* on which the musical play and movie *My Fair Lady* were later based. The theme dealt with the English language. Sixty-six years ago, in his will, he wisely left funds to sponsor a phonetic alphabet. A true phonetic alphabet will not include now silent letters like the "*k*" in knife and knight, and the "*w*" in write, and wrong. It will not have the additional spelling for the "*f*" sound as in the "*ph*" combination in photo and phone. Every symbol will correspond to a single sound including sounds not in or rarely heard in English like the different "c" sounds like in cough and *mach* referring to the speed of sound as *mach 3* meaning three times the speed of sound. Also the Dutch "*ch*" in *Scheveningen*, the beach resort of The Hague in the Netherlands and the Hebrew word *chaim meaning health and life.*

A phonetic alphabet will help stabilize language speaking and writing. It will greatly reduce the time now wasted in looking up spellings and meanings in dictionaries and on the internet. Then, if you can pronounce it, you can spell it. This will not happen quickly but hopefully will eventually happen. The adoption of a phonetic alphabet will be of greatest benefit to new immigrants and disadvantaged students. This may be politically helpful in getting funds for the difficult transition. A phonetic alphabet could be used worldwide to replace the spelling of individual languages.

CHAPTER 1

∞

Memory – the Importance of Forgetting

In my working days I frequently forgot where I had parked in the office parking lot that morning, instead remembering where I had parked the day before. At the 1965–66 New York World's Fair, there were 15 parking lots with shuttle buses to the fair. As I returned to a parking lot, I saw people who had forgotten which parking lot they had used.

Being able to forget information is useful. There are people called savants whose memories are saturated because they cannot forget. They are unable to function normally. There are also savants who are saved from being geniuses by lacking common sense and influence others who are vulnerable.

I have a very erratic memory. Despite repeating a name to myself silently when I am introduced, I can quickly forget it. I would be rotten as a politician. I can sometimes recognize a face and know much about a person, but not recall the name. Now we know that different types of information are stored in different parts of the brain. By contrast, politician Jim Farley, who helped Roosevelt get elected, remembered faces of thousands of people he had met years before and connected those faces with their names.

But I can remember some things from most of a lifetime ago; some examples.

My earliest memory is from when I was about a year old lying on my back in a baby carriage sweating, bundled up by my protective mother to keep me warm on a hot day and under mosquito netting cutting down the air circulation. I heard noises that a year or so later I interpreted as iron horseshoes on a cobblestone street and the steel rims of horse-drawn wagons on that street.

I remember standing in my crib screaming my head off for my mother who was someplace else while my dad was trying to calm me down; another time my dad placing me on his shoulders saying "little me." I remember when I was two years old sitting in a chair with cut out ducks. I was facing north, parallel to the ocean and boardwalk in Atlantic City and viewing a fire escape; thus memory of rooms and directions.

Before supermarkets, in good weather small stores placed boxes of fruits and vegetables outside under awnings. When I was about four or five, as a passenger in our car driving by the store, I saw an elongated watermelon and trying to be cute said, "Look at the big cucumber." I failed to convince my parents that I actually knew the difference.

A memory of mine from 1939 is when I was in high school. My father was running his photography business in Hollywood, Florida where I went to school. I was taking my first science course, biology. My teacher introduced himself as Charlie Perkins. At my present age it took me some time to remember his last name. This teacher was kind enough to allow me to read his science magazines while he lectured, as long as I was able to answer his questions when called on.

About six weeks before the end of the school year, Charlie Perkins had to take over for the only other male teacher in the school to teach wood shop. I replaced him as the biology teacher. In my yearbook I was called "Professor." I have remained very interested in biology.

During World War II, I would frequently hear an annoying ad on the radio urging, "Rush, rush, rush for Orange Crush," (a soft drink). I would have boycotted it.

During the academic year 1946–47, I was a university physics teaching assistant in greater Los Angeles. The major soap companies were advertising that their soaps made more bubbles in washing machines. I remember an ad on a wall above the windows of a streetcar by a local area soap company saying, "How to get twice as many bubbles with Stryker's Soap. It's ridiculously simple. Use twice as much soap." Later I learned bubbles actually interfered with washing laundry.

The university professor supervising me told me one of my 100 plus students complained that on the streets of Los Angeles I had walked by him without saying hello. I explained that with a 100 plus students a week I had not recognized him off campus.

Specific parts of the brain have specific functions. This was recognized during World War I, as physicians treating soldiers wounded in their heads by bullets or shrapnel observed what capabilities had been deleted by the damage to particular locations, for example, the ability to sleep. Sometimes, as after loss of eyesight, the part of the brain processing images may reprogram itself. Of course, the path of a bullet or shrapnel through the brain raises havoc along its whole path, doing much more damage. The paths in other patients who also lost the ability to sleep but had a different damage path through the brain were compared. The sleep center is at the intersection of these paths. Currently, with Functional MRIs, it has been verified that different types of information are stored in different parts of the brain. It has been shown that very young children who learn multiple languages store them all in one part of the brain. Later on in life, a different part of the brain is used to store new languages, making them more difficult to learn.

What is called instinct is inherited memory. A newborn infant suckling on its mother's breast is one example. A whale giving birth at sea and its newborn coming to the surface to take deep breaths and then suckling on its mother's breasts is another. A colt standing on wobbly feet immediately after leaving its mother's birth canal is still another.

The often quoted supreme example of inherited memory is the migration of the monarch butterflies. Starting in New England or Canada, stopping in the Carolinas where they lay their eggs, the eggs maturing into caterpillars, the caterpillars dining on leaves and then molting into the next generation which continues the flights of its parents to a particular forest on a hillside in Mexico where it lays its eggs to grow into caterpillars to feed and molt becoming adult monarch butterflies which begin their journey northward to where their ancestors came from.

In 1947, to earn a living, I went to work for the Naval Research Laboratory (NRL). To illustrate my erratic memory which often remembers details from long ago better than from yesterday, I will state some selected details from my employment there.

During the 1920s, the propagation of radio waves was researched there; this means how radio waves traveled including bouncing off of objects. This led to the development of radar for navy applications. The army was also developing radar at Fort Monmouth, New Jersey. Massachusetts Institute of Technology (MIT) was participating. In the United Kingdom, radar was being developed, expedited in preparation for World War II, especially after the 1938 Munich agreement which sacrificed Czechoslovakia to buy time which allowed the British to complete a radar/command and control system to see approaching German bombers and their escorting fighters and to direct interceptors to shoot them down.

I heard that during World War II the navy had trouble with draft boards that were interfering with the war effort by drafting engineers and scientists from NRL. The navy resolved this problem by enlisting them as navy officers to continue their technical work at NRL.

There were many complaints about the food so the cafeteria was redecorated, including colored banners hanging from the ceiling, to provide a better atmosphere; but the food remained the same.

NRL had 2,000 plus employees. It was bordered by a high fence and the shore of the Potomac River with a short pier protruding into the river. While employed there, bordering the downstream side of the fence, was a Washington, DC sewage treatment plant which I could smell when the wind came from that direction. There were many buildings and parking places.

So many years later, I still remember the name of my first boss at NRL. If he is still alive he would be in his mid-90s. The name of my second boss, I have forgotten.

While I was working for NRL, there was an attempt to requisition electric fans to help make the hot, muggy summers tolerable; air conditioners were ruled out because navy field locations could not have them. There was an audit to determine how many electric fans had been procured since NRL was founded and how many were on the premises. About 2,000 were missing. Security was tightened up. All employees wore badges with photos and names on them. Security guards looked at them when we entered the gate and when we left.

The director of NRL was proud of the improved security. An analyst friend working for him challenged his confidence in the security. He told the director to take an unclassified document and label it with the highest security classification. (I was not yet familiar with the preparation of classified documents; a few years later I was writing many classified documents so the description here is based on my knowledge later.) He would smuggle it out and invited the director to visit him in 30 days and he would hand it to him. The director required the security officers to very thoroughly search the analyst and his car. The director went to the analyst's home and was handed the document. The analyst then told him how he had done it. He had simply taken a very large envelope, printed his own home address on it, placed the document inside, sealed it, handed it to his secretary, and told her to take it to the mail room and mail it.

CHAPTER 2

∞

Aging – Your Health

Throughout my life, I have taken responsibility for my health by researching medical issues in order to be able to advocate for my health with my medical professionals.

A major issue today is inoculations causing overstimulation of the immune system. Inoculations can damage your children's health even as adults. Many cases of ALS are probably caused by this overstimulation. Children should be inoculated for one disease at a time instead of receiving multiple inoculations for the convenience of physicians.

People are interested in why and how we age—a significant health related fear is dementia. There are several types of dementia. You can have more than one at the same time. I will now describe them and what you can do to try to prevent them.

The first type of dementia is trauma-related. Contact sports like soccer and American football as well as combat related injuries caused as by improvised explosive devices

(IEDs) are examples. Boxing has long been recognized as leading to brain damage and thus dementia. High schools and colleges are responsible for damaging the brains they are trying to educate—a conflict of interests. To reduce accident related brain damage, be a safe driver. Wait until icy roads have been cleared before you drive on them. Do not drive under the influence of alcohol or drugs or when you might fall asleep at the wheel.

Second, there is vascular dementia caused by clogging of the blood vessels so not enough oxygen reaches the brain. The clogging is the result of too much fat in the diet being deposited on the walls and within the walls of the blood vessels. The blood being pumped by the heart divides into two parts. One part goes to most of the body including the brain. The other part goes to the lungs to absorb oxygen from the air in the lungs. Smokers coat the walls of their lungs with tar which reduces the amount of oxygen in the blood returning to the heart to be pumped around the body. Smokers often develop pulmonary hypertension which is very high blood pressure in the lungs which can be far greater than that measured with a cuff on an arm. Smokers with clogged lungs often are pulling an oxygen tank on wheels so they can breathe more oxygen and extend their life a few more years.

Third, Alzheimer's disease, named after the doctor who defined it. Holes develop in the brain where nerves have died and the body defends itself by a kind of glue to keep the brain from crumbling; that glue is named amyloid protein. Recently brain tissue from autopsies of Alzheimer's patients was genetically tested and found to contain DNA of common fungus as in mold in the walls of a damp building with a leaky roof.

All animals, including us, are more closely related to fungi (mold, etc.) than to bacteria and insects. Anti-fungal medications are harming us, destroying our livers. To me, the work against amyloid protein by pharmaceutical (drug) companies is "barking up the wrong tree."

I have observed that as people become demented they sometimes become happy because they have forgotten what they were upset about.

Next, a little biology relevant to aging. A normal person has 23 pairs of chromosomes in the nucleus of every cell. The function of a cell—what kind of tissue it is in, nerve, bone, muscle, etc., is determined by which chromosomes and parts of chromosomes are turned on.

In each cell outside the nucleus there is also other inherited material. Of particular importance are the mitochondria which determine your energy level, the metabolism of the cell, burning food with oxygen such as starch, sugar and if necessary protein. When you are running fast or doing other extreme exercise you are going beyond the ability of your normal metabolism to keep up and your body resorts to anaerobic (meaning without using oxygen) metabolism which produces lactic acid. This causes fatigue and probably causes mitochondria precursors to become functioning mitochondria. When that happens, it can disrupt the body's production of chemicals for the body's cells to communicate with each other which signifies cancer. If this situation lasts long, it is cancer.

Since a female is born with all the eggs she will ever have, aging begins before birth and is also influenced by the female's mother and her mother's mother. Soon after World War II, I read an article and because it was interesting I remember the key points. Norwegian investigators went to fishing villages at fiords that had been isolated for centuries. They examined church records. They calculated life spans of members of the same families. They determined that (for the same sex) the life span of the first born was longer than the next born by about half the difference in ages at death. This means aging of the mother's eggs before becoming fertilized was continuing between when the mother was conceived and when she herself conceived a child. It is likely that the age of the grandmother when the mother was conceived also influenced the life span of the grandchild (my informed speculation).

Why does hair turn grey? Skin and nerves are being formed in the outer layer of the embryo as the fetus is being formed. Skin, nerves, and hair color all use melanin. As we age, our body can no longer produce the melanin that we need. The melanin that goes to the nervous system takes priority over the melanin that retains our hair color.

White skin has as much melanin as black or brown skin. The difference is that the length of the melanin pigment in skin is tuned to the wavelength of light, just as antennas are tuned to the wavelength of the radio waves/microwaves. Skin color is a balance between getting skin cancer and avoiding rickets and having strong bones. It goes back to the human hunters who chased animals until the animals dropped from heat exhaustion. Animals with fur covering their skin have the capability to produce vitamin D without sunlight; humans have lost it. Use it or lose it.

Why do humans live about seven times as long as dogs? Why do large dogs tend to live shorter lives than small? Why do mice live shorter lives than dogs? Why do women now tend to live longer than men and about 70% of those reaching 100 are women? Why do baseball players tend to have longer lives than the generally larger/heavier football players? The answer long ago provided by biology professor, Lynn Margolis, was that evolution on earth included the merging of two different species (called symbiosis). Some genes from mitochondria migrated to the cell nucleus providing better control of cell reproduction and growth.

Another example of symbiosis is vital to the lives of mankind. We depend directly or indirectly on vegetation for our energy. We receive energy from plants which receive energy from the sun. Plant cells which we call chloroplasts include both mitochondria and algae.

Within a species, the reproduction of the mitochondria within cells is mismatched to the reproduction of the nucleus of the cells. A longer pregnancy allows more time for mitochondria to reproduce. In nature, a balance is struck between reproduction in a life span before being eaten or otherwise dying, and natural life span as limited by aging which depends upon the time of mitochondria reproduction. Within a species, larger bodies tend to burn more calories. Therefore, they age faster and die younger.

Aging is a defense against cancer. Every cell in our body has telomeres. These limit the number of times a cell can divide over a lifetime. Telomeres eventually stop cancer cells from being able to continue to multiply.

Tree galls resemble cancer. There is a suspicion that something in a tree gall can spread to humans in nearby homes and induce cancer. The frequency of disease is more than a coincidence. It could be an infection that causes some cancers.

To lighten up this otherwise gloomy chapter I insert a relevant humorous anonymously written poem on aging I had read perhaps a half century ago and now found on the internet:

> The horse and mule live thirty years
> And nothing knows of wines and beers;
> The goat and sheep at twenty die,
> With never a taste of scotch or rye;
> The cow drinks water by the ton,
> And at eighteen is mostly done.
> Without the aid of rum or gin
> The dog at fifteen cashes in;
> The cat in milk and water soaks,
> And then at twelve years old it croaks;
> The modest, sober, bone-dry hen

> Lays eggs for nogs and dies at ten;
> All animals are strictly dry;
> They sinless live and swiftly die,
> While sinful, gleeful, rum-soaked men
> Survive for three score years and ten.
> *(Author Unknown)*

A score means 20, so three score and ten years means 70. I have turned 90, eight score and ten, and am still both alive and functioning. My darling wife is a little younger than me—still driving and hosting family and friends for dinners and teas. The older I become while still functioning, the more credible the health advice I am giving here, with perhaps an updated edition of this book at 100, if I make it still functioning, it will be even more credible.

Cell constancy

A single sexually conceived cell continues to divide until it has 959 cells in the species of the tiny earthworm called C. Elegans. Likewise, the count of cells in a part of the kidney in mammals has been found to be the same for individual mammals within a species. This is known as cell constancy. You were born with all the nerve and muscle cells you will ever have. Your growth after you were born was only by growth of your cells themselves. The only cells that multiply are cancer cells and those that wear out like skin cells, hair, the lining of your digestive system and blood cells. The life of a red blood cell is about six weeks which is why the recovery time from most things is six weeks.

The jaw of the salmon

Salmon on the west coast of North America go to sea and after they are fully grown navigate, using their sense of smell, back to the stream and headwaters they were hatched in. Since this began, the Pacific coast mountain range climbed higher and higher as the continental shelf slowly moved out over the sea floor. Due to the additional exertion needed to handle the incline, the salmon's body changes. As the salmon begin to leave the sea, their jaws grow and become distorted as in the human disease acromegaly. About 1968 I drove my family up to Nova Scotia on a vacation. In a museum at Halifax I saw skeletons of two giants, one male, one female, whose jaws and hands as adults had grown and distorted. They had died young. The cause was excess release of growth hormone from their pituitary glands due to cancer. The jaws of the salmon had likewise been stimulated to grow by a great dose of growth hormone. After the female salmon lay their eggs on the floor of the headwaters stream and the males fertilize them, the salmon lose their energy and within weeks become listless and die of old age. In contrast, Atlantic salmon go out to sea again and, if surviving, may repeat the cycle. Humans who take growth hormone to help create bulging muscles are likewise speeding their own aging.

Polar bodies

A cell to become a normal egg must get rid of half its chromosomes at random, selecting half inherited from the female's mother and half from the female's father. The mitochondria normally remain in the egg. The remaining chromosomes are ejected in what is called the polar body which normally deteriorates.

What if, because of inheritance of normally recessive mutations from both parents, some of the mitochondria go into the polar body before it deteriorates and the polar body survives along with the normal egg? This would be a very unusual happening. The polar body could act like an egg and be fertilized. A child is born; it is doomed to extreme early aging, usually dying of old age by about 20; this disease is called progeria. About a half century ago, I wrote a paper describing aging including my original explanation of progeria and submitted it for publication. It was rejected. In rejections, I have good company, including many other scientists.

Health Tips

For your own health, tobacco is a strongly addictive gateway drug that by custom since early colonial times is legal, grandfathered in. Relatively few nonsmokers become addicts to anything. It also became a major export, important to the economies of Virginia and North Carolina. Tobacco/nicotine causes the early death of more people in the United States than all other drugs combined. It is also a gateway to other drugs. Those who do smoke are setting a horrible example for their children who are more likely to become drug addicts. As a young boy I was turned off on smoking by visiting uncles who stunk up our apartment by smoking. I have never smoked, not even once. But I have been among the victims of second-hand smoke in work places when that was legal.

While Margaret Thatcher, herself a smoker, was prime minister of Great Britain, the British government declined to have a vigorous antismoking campaign because it was concerned that it would not be able to finance peoples' pensions and health care if they lived much longer.

About 65 years ago, I had a young co-worker who smoked a few packs a day. He rationalized his addiction by saying he did not want to be in a nursing home in his old age. I lost track of him. He probably is long gone, possibly by way of sooner entry into a nursing home on the way.

In the 1930s when I was a young boy we were eating organic food and didn't know it. Pesticides and weed killers had not yet been developed. They are what I call biocides (life killers). Biocides are often prescribed by physicians. We are more closely related to fungi like athletes' foot, toadstools, and mushrooms than to bacteria. Drugs that are effective against fungi are also effective against us, sometimes destroying our liver and killing us.

One problem was trans-fats. Vegetable oils were being hydrogenated to make them solid grease for cooking/frying substituting for butter. People were saying they were made from vegetable oils and were therefore healthy. Arsenic and mercury are also natural; natural can be poisonous.

Lunches I brought to school included a small bag of raisins and walnuts plus a peanut butter and jelly sandwich. The oil in the peanut butter was not hydrogenated; it separated out, was stirred in, and was healthy. Later, the oil was hydrogenated to solidify for lazy customers.

Today there are reasons for suspicions that polyunsaturated oils like corn and safflower oil, customarily considered very healthy, contribute to heart attacks and strokes.

Glycemic index is a relative listing of how quickly a food is converted in the body into blood glucose which the body burns to produce energy. About the worst is high fructose corn syrup, frequently in cans of fruit. Starch has a higher glycemic index than sugar and is more rapidly converted to blood sugar than sucrose from sugar cane and sugar beets.

I avoid artificial sweeteners which I suspect are harmful. One was actually developed as an insecticide. A researcher accidentally got some on his hand and instinctively licked it, discovering it was sweet, so the company marketed it, instead, as a sweetener which I see among other artificial sweeteners on restaurant tables where coffee and tea are served. It is also sometimes an ingredient in vitamin B-12 and perhaps other pills.

Too high blood glucose makes type 2 diabetes (often called adult-onset) worse. In type 2, the body has poor response to injected insulin or insulin produced by the pancreas. Spikes of insulin in the blood are damaging, often resulting in heart attacks or strokes. My father took a medication prescribed to stimulate the body's insulin production. I believe it killed him at age 84; dying of a stroke. Small meals with snacks between meals help avoid insulin spikes.

The usual diet will contain sufficient Omega 6 oil. Omega 3 oil as in walnuts, flaxseed, and wild sea fish is healthy. Omega 3 oil originates in one cell plants (called algae) living in the sea, and is in the marine life that eats them. A good source is canned sardines in olive oil. Avoid large fish high in the food chain that accumulate mercury. Thus avoid tuna, sea bass, halibut, and swordfish.

For a healthy diet, follow the common advice to eat a rainbow of colored fruits and vegetables. Raw is good but sometimes, as with tomatoes which contain lycopene, cooked is better. The lycopene in cooked tomatoes has been shown in a study to be highly correlated with a lower rate of prostate cancer—probably good in preventing/delaying other cancers as well. President Reagan was ridiculed but was right in stating that ketchup in school cafeterias should be considered a vegetable.

Cinnamon is made from the bark of a tree. It helps protect the tree from fungus and bacteria. It does the same for you. In the days before refrigeration cinnamon and other spices were used to help preserve food.

When your blood glucose (the form of sugar your body burns for energy) is low you become hungry. To lose or avoid gaining weight an effective diet is to eat dessert first, then wait 15 or more minutes while your blood glucose rises, spoiling your appetite so that overall you eat less.

Don't wreck yourself for your health. Marathon runners are wearing out the cartilage of their joints and often have knee and/or hip joint replacements and spine damage. Later, because of pain, they become couch potatoes. Data on life spans indicates that they live no longer than other couch potatoes. Moderation is the key. Walk, don't run, don't jog. Avoid jumping and other impact exercises.

The purpose of pain is to stop you from doing whatever you are doing so as not to damage your body. I do not use pain killers except a little alcohol for a toothache until I can get to a dentist. Magnesium can prevent or relieve a headache.

Don't be skinny. Women who are skinny have more difficulty becoming pregnant and have an earlier menopause. Statistics for nonsmokers show that as we become older a little extra weight, an energy reserve, helps us live longer. When I was around 60, my pants-size waistline was 36 inches (91.4 centimeters). Now at 90, it is 38 inches (96.5 centimeters). I am seeking to keep it there.

I have a balance problem. A nephew, the son of a sister of my wife, who is a research biochemist/cardiologist advised me to ignore the previous standard advice and eat eggs. He said the vitamin B-12 in eggs is good for balance and birds had evolved so that both the high density (HDL) and low density (LDL) cholesterol were good for the chicks that hatched out and were good for you. This advice corresponds to common sense. I followed his advice and within a few months found a big improvement in my balance. Stick to eggs, fish, poultry and meat as your source of B-12. Avoid a vegan diet; you get no B-12 from plants. Beware of vitamin B-12 in vitamin pills. On the label of a bottle of multivitamins I read vitamin B-12 was listed as cyanocobalamin. This is cyanide, a poison, linked to vitamin B-12. Cyanide accumulates in your body.

Genetically modified (GMO) plants are sometimes good, sometimes bad. Taking genes from lobsters and putting them in plants to be eaten to improve a plant's frost resistance was a rotten idea and was not approved by the American Food and Drug Administration (FDA). Taking genes from marigolds and inserting them in rice was done, eliminating a beta-carotene deficiency among rice eaters in Asia—a good idea. The body converts beta-carotene to vitamin A as needed, avoiding the harm done by excess vitamin A.

To avoid becoming obese and to avoid insulin spikes, eat no big meals. You can make an exception on rare occasions such as Thanksgiving. *Nosh*, eat a number of times during a day, like five. If you get up late, as I do during retirement, combine breakfast and lunch as brunch. Similarly, late in the afternoon, too early for supper, you can combine them into what I call lupper. Then before going to bed, *nosh* again with another small meal which you can call pre-sleep or whatever you choose. The important thing is preventing your blood glucose level from going too low or too high. Recently when I get up a few hours before breakfast, I eat organic dried figs or organic prunes and drink a glass of water. This helps stabilize my blood glucose and I eat a smaller breakfast. The word breakfast, which we take for granted, means breaking the overnight fast.

At 90, I find myself having difficulty doing things I took for granted when I was younger like buttoning the sleeves on my shirts. As you age, keep doing things you can do for yourself. "Use it or lose it." There are exceptions for safety. I stopped climbing a ladder to do maintenance on the roof of our house for safety reasons. After a stroke, I stopped driving at 89.

Passing kidney stones is extremely painful. The doctor coming to our home where my late mother-in-law was visiting as she passed a kidney stone told me he had arrived too early; she might have given him credit if he had arrived later as the kidney stone had passed and the pain was subsiding. Standard medical advice is to drink lots of water and other fluids but is often incomplete. Kidney stones are formed when urine is very concentrated, a deep yellow. Before going to bed drink enough water or tea so you will have to get up at night. Then when awake, drink more fluid so you will have to get up again and your urine never becomes concentrated and you will never grow kidney stones.

If you sometimes get a little dizzy when you stand up quickly that is a sign that your blood pressure at that moment is too low. Before you stand up, raise your blood pressure by waving your arms and kicking your feet. If you idolize low blood pressure, feel consoled by the fact that when you are dead your blood pressure will be extremely low, zero. When you are measuring your blood pressure first sit or lie down for at least five minutes. Your blood pressure goes up when you are active, which is normal.

As I was beginning to grow facial hair and showing other signs of puberty, my father lectured me sternly on the dangers of sexually transmitted diseases. I was already aware of an uncle who was sterile, unable to father children of his own. I took my father's warning very seriously. Later I became aware of other relatives becoming sterile. Decades later, AIDS was found among male homosexuals in the United States, having been brought from Africa. Some who had AIDS were bisexual and gave it to their wives and girlfriends.

The only person I have ever gone to bed with is my darling wife (of 66 years, as I write this), the mother of our children, the grandmother of our grandchildren, the great-grandmother of three great-grandchildren, so far. Still beautiful in my eyes, aging gracefully, showing our sons-in-law what their beautiful wives will look like at various ages.

A warning. For some people, a flu shot can be very dangerous. Taking an aspirin with a virus in your body such as flu, but also other viruses, can cause paralysis. Reyes Syndrome is caused by taking an aspirin while you have a virus in your body.

A college friend of my daughter drove up to our home about two years after graduation and took a walker out of her car. She needed it because she was partially paralyzed. As a nursing student, she was required to take the Swine Flu shot. She might have had mononucleosis at the time.

Also a relative got the flu and became very sick after taking a flu shot. Probably his resistance was lowered by the shot and the shot was against a different strain of flu virus, selected before the flu season.

My wife and I have never had the flu and never had a flu shot. Unfortunately, our children did not inherit our natural immunity.

A cold is appropriately named. When your body temperature drops, you are more susceptible to germs. It was also observed that fish being raised in cold water were more apt to die than if the water was warmer. You lose about a quarter of your body heat through your head when you are outside in cold weather—wear a warm hat, and inside in a cold room also wear a hat and a jacket or a vest.

Chickenpox is caused by a virus, usually in childhood. The virus remains in the nervous system for life and often causes shingles in older adults, with painful irritations of the skin, often lasting for months. In the human embryo both the nervous system and the skin form in the outer layer so it should be no surprise that a virus in nerves can also affect the skin.

Vaccination for multiple diseases at the same time can overstimulate the immune system and lead to recognized and not recognized autoimmune diseases. Combining vaccinations is easier for parents and physicians and other medical personnel and helps assure that all the multiple diseases are vaccinated for; but can place time bombs in the body that may detonate later, perhaps as ALS/Lou Gehrig's Disease, named after a then famous baseball player instead of a physician (descendants of Dr. Alzheimer must have changed their name). Whether or not you will suffer from Alzheimer's disease may depend upon whether your immune system was overstimulated as a child. So-called experts in and out of government cannot know the full long term consequences of what they recommend.

Very good friends had a young grandson who was developing normally until he was vaccinated as required before registering to go to school. He became autistic which so-called experts and authorities dismiss as coincidence. A pedestrian crossing a street against a red light and getting run over is likewise coincidence.

During the Korean War, I had a coworker at the Naval Gun Factory in Washington. He lived with his wife and two young children in the same new apartment project as we did. He had a stomach ulcer. Many years later I realized that it was probably induced by an antibiotic which killed off both the helpful and the toxic bacteria in the digestive system, allowing the toxic bacteria to take over. Specifically, H. pylori bacteria penetrated the wall of his stomach creating an ulcer which over years of inflaming the raw wall of the ulcer causes stomach cancer. To prevent this, eat yogurt with multiple live cultures at least once a day.

The greatest enemy of your health is stress. Stress increases the flow of the hormone adrenalin, a stimulant which readies you for physical combat. To reduce stress, work off your adrenalin overload by taking a regularly scheduled walk, for instance, during lunch break, or do other physical activity in moderation. Try to reduce stress by relaxing and meditating. Get enough sleep—take naps when you can. A half-hour nap is worth many times an extra half-hour's sleep at night. On reducing stress, advice often given is, "The best thing a father can do for his children is love their mother and keep her happy."

Why African Eve?

African Eve lived in Africa over 60,000 to 100,000 years ago. The first Africans out of Africa were Bushmen like those now in parts of southern Africa inland north of South Africa. The estimate depends on the mutation rate of African Eve's mitochondria. The original mutation gave her descendants their great evolutionary advantage of longer life span so they could help raise their grandchildren and thus, themselves, have more grandchildren. Bushmen were then living in the horn of Africa at the opening from the Indian Ocean into the Red Sea. The climate was very different then with much of Europe covered by glaciers and heavy rainfall in what is now the deserts of Arabia and Africa.

African Eve may have been a female Bushman or a descendent of others who mated with those who had the mitochondria for long life span. African Eve's descendants mated with pre-existing very diverse populations outside of Africa and her mitochondria genes gradually took over. There are more genetic differences within the native African population than between Africans and other populations.

In apes other than humans, menstruation and the ability to give birth continues through their shorter life spans.

Mitochondria of Neanderthals was extracted from bones, analyzed, and found to be different. Neanderthals had larger brains than Homo sapiens. They were far from stupid. Fossils show they made a special glue from the bark of a particular tree by heating it in a very hot fire and then using it to glue sharpened stone points to wood shafts to form spears and arrows. They were better adapted to Ice Age climate with stocky bodies and shorter arms and legs to conserve body heat but in demographics over thousands of years were beaten out by the descendants of African Eve who, with their longer life spans, could help raise their grandchildren. Neanderthals who mated with Homo sapiens and humans, other than Africans, have some Neanderthal genes.

Humans live longer than other apes. The genes of those whom African Eve's descendants interbred with predominated. Genes from the cell nucleus of a Neanderthal finger bone were later also recovered. That particular Neanderthal had red hair.

As with a three dimensional jigsaw puzzle, that may be why there are individuals born with red hair.

In genetic testing of my saliva, which has become inexpensive, my heredity was found to be 2.6% Neanderthal. Genes for red hair are probably linked to other genes, as letters in a crossword puzzle are in both vertical and horizontal words. Genes for red hair may have been linked to advantageous genes for immunity. The analysis also showed Middle Eastern and northern European heredity which I already knew. I have blue eyes and my hair was blond before it turned grey. By the traditional way of determining heredity, I am like Moses, a descendant of Levi, the third son of the biblical Lea and Jacob (also known by the name, Israel) roughly 4,000 years ago.

My own aging

As a teenager I sometimes rode my bicycle around corners while leaning with my hands off the handlebars while gripping the frame. In my mid-70s, after a winter of no biking, I put my helmet on, got on my bike and felt insecure, feeling I might fall down. I quit bike riding right there. By my early 80s I was using a cane to help balance when walking outside where there might be some rough ground or pavement to trip over. Now, 90 as I edit this, I have graduated to a walker with wheels and brakes—outdoors and indoors—most of the time, and the rest of the time mostly holding on to furniture and counters. Also in my 80s, going through contortions to cut my toenails became difficult. My vision is better now than it was before lens implants replacing cataracts as I was turning 60—too much bright sunlight without sunglasses in my youth. I now take eye drops to prevent damage from glaucoma. I recently had my ophthalmologist check my vision including field of view and it was about the same as ten years earlier. My erratic memory is nearly the same as it has ever been, allowing me to write this book. But it often takes me longer to recall a name or fact.

To cheer you up about aging I end with a humorous story I heard recently. An elderly parent being helped with a computer by a grown-up child responded with, "Remember how I taught you how to use a spoon."

CHAPTER 3

∞

The Creation of Time

The Bible Ecclesiastes: "There is a time to be born, a time to live. And a time to die," I add this also applies to the universe we live in.

The first three words of the Bible translated from Hebrew are "in the beginning," thus assuming creation—not that this universe has always existed. The first chapter of the Bible is named "Genesis" meaning origin, creation.

The dark night sky

Centuries ago it was realized that if the universe we are in extended forever (and was not expanding) between every star we see there would be other more distant stars and the whole sky would be as bright as the surface of the sun. This universe would be as hot as the surface of the sun. Life could not exist. This was stated by Oblers, and others long before him, and is known as Oblers' paradox. This idea that the universe is expanding was observed by seeing that the color of spectral lines of stars shifted more the farther away the star was. The color shift is known as the Doppler effect. This is the generally accepted interpretation. But the *Astrophysical Journal* in 1979 reported "enormous periodic Doppler shifts in ss 433." I interpret this as enormous oscillation in size as toward and away from becoming a black hole, not a change in distance. And/or what has been treated by conventional astrophysics is wrong in determining distance to, and the age/maturity of, galaxies containing quasars.

Pointing the arrow of time

What has been called a crisis in physics was created by the standard model which uses mathematics which have time symmetry, meaning not distinguishing time going forward and time going backward. Thus the standard model fails the test of reality. The standard model in theory requires the existence of the Higgs Boson, a particle that is part of larger particles like protons and neutrons and gives them mass, otherwise we could not exist.

First a brief digression to help illustrate why an infinitesimal (very tiny, tiny, tiny, tiny) difference can help explain why time can only go forward. Time machines are impossible. We are not overrun with visitors from the future changing our present and their past. Some science fiction stories and movies (such as *The Terminator*) include going back from our future to change their present. In one story, someone goes back carefully but steps on a butterfly. He returns to his present and finds a different language pronunciation and a different government—"the butterfly effect."

The British author physician, Conan Doyle, wrote the Sherlock Holmes detective stories. In one story the clue was that the dog did not bark, meaning that the dog knew the murderer. Sometimes the absence of something is the clue.

In nature, not everything acts the way you might predict. For example, there are no tornadoes or hurricanes (also called typhoons and cyclones) near the equator. In the northern hemisphere tornadoes and hurricanes swirl counterclockwise, in the southern hemisphere clockwise.

The equations that physicists had been using do not distinguish between time going forward and time going backward. With the universe expanding at an accelerating rate, energy going forward in time is infinitesimally smaller than if time were to go backward. The universe is rapidly growing, with time only moving forward at a faster and faster rate. Thus, pointing the arrow of time.

In the 1960s I wrote various versions of a paper, "The Creation of Time." I tried to reconcile that the universe we live in has existed forever and the idea that before the big bang it was extremely tiny. Time before the big bang correspondingly crawled along at trillions of trillions of trillions of times slower than now. Time has existed forever. The universe has existed forever, but it had been trillions of trillions of trillions of times smaller. Existence implies time.

Modern experiments with the latest technology show us that theoretical equations did not make correct predictions.

Unfortunately, I did not date the various versions of my paper. Fortunately, copies of a relatively short late version of my paper were returned from *Nature* with a rejection letter dated 23 January 1976 stating "we found it too speculative for inclusion in *Nature*."

"The Creation of Time"

A cosmic clock going faster and faster is equivalent to the Doppler effect in accounting for the redshift of light from distant galaxies. Implications include:

With theories of the creation of time, space, and matter there is time asymmetry and a preferred inertial frame. Accelerating expansion of the universe is predicted.

The clock going faster and faster has given a relatively greater radial expansion to space near us, which is transited by the light we see at closer to our present time and clock rate, than to more distant space. Therefore, the apparent density of galaxies increases exponentially as we look farther into space.

Apparent radial compression of the space between galaxies is proportional to the ratio of the emitted radiation to the observed wavelengths of distant galaxies.

Energy of radiation is proportional to frequency. The less time required for an event, the greater the energy. As we look into the past we see photons with less and less energy—the redshift. In an extreme example, measured in terms of today's clock, photon energy was zero and mass, being equivalent to energy, was likewise zero. This is with measurements performed now. Looking into the past, distance is measured with an elastic ruler. The result would be the same for an observer at any time in the past or future, and for any place. To each such observer in his own time the universe and its gross features would appear much the same, independent of the observer's epoch—provided however that the average density of the universe remains the same, which may not be the case.

The "big bang" theory is that the universe exploded in about a trillionth of a trillionth of a second from a miniscule point. I will not attempt to explain the cases for or against the theory of inflation which was most recently explained by Alan Guth in 1980, building upon the work of other physicists over many years before him. The extremely tiny differences in background microwave radiation detected from different directions in space are evidence of inflation predicted by physicist George Gamow as radiation from the "big bang" was stretched to microwave lengths as the universe expanded; this is the reverse of the infinite time it takes to complete falling to the center of a black hole. The further you fall into a black hole, the slower time is passing, so you would never reach the center. I am very skeptical of the theory of inflation.

It is more likely that the universe has always existed, that it is infinitely old and that, as measured by our present clocks, time crawled—trillions times trillions of times slower.

Life is only possible in the slice of time, measured in billions of years, between the explosions of quasars and supernova. Hydrogen and helium which existed in the quasars and supernova were the raw material from which heavier atoms were created, such as oxygen, carbon, sodium, magnesium, calcium, phosphorus and iron, etc. These are necessary for life. Life will end in the universe as it flies apart over trillions of years. All atoms will disintegrate. All the stars will have used up their fuel and be burned out. The universe will be dark.

Alfred Nobel, the inventor of dynamite, established funding for the Nobel Prizes in his will. The Nobel Prizes, beginning in 1902, included a prize for physics. Perhaps it is especially appropriate as the universe is blowing apart.

The cockroach theory of cosmology

If you see one cockroach there are more. We see one universe, the one we live in. I am speculating that there are more, an infinite number. Even if the number is infinite there is a subset that is infinite, and like ours, includes places able to evolve life and civilizations. Each universe has its own space-time and we can never have had contact with it or it with us. This is what was once called natural philosophy. What is now called science demands that experiments be repeatable by other experimenters—impossible with this speculative theory.

Creation of matter

If matter is not continuously created as the universe expands, as postulated by Hoyle, Bondi, and Gold, the universe would proceed toward becoming a complete vacuum, but would never actually get there. The existence of space seems to be linked to the existence of matter and conversely the creation of space is expanding the universe which in turn is linked to the clock going faster and faster and faster. Thus there is positive feedback, that is a bootstrap operation. This is indicated by the bending of starlight by the gravitational field of the sun. Assuming that the "big bang" created matter is hardly satisfactory, for how was the big bang initiated? A mechanism is needed to explain, rather than postulate, the creation of matter.

I had once read in the vast writings of Moses Maimonides (1138–1204) that time proceeds in very small steps. If time proceeds in very small steps, then it follows that space, charge, and mass are likewise quantized—the same size building blocks. Otherwise how does one electron "know" to have the same mass and charge as another? The steps of time thus scale the universe. (I add here that the quantization of time and space is also the reason of the mass and charge of all particles.)

At this point I am briefly getting off the subject to write about Moses Maimonides who is very interesting. He was born in Andalusia, which was the Muslim ruled region of most of what is now Spain and Portugal, conquered by North Africans who had been converted to Islam by missionaries from Damascus, and taught Arabic so they could read the Koran. They ruled Andalusia for eight centuries until they were finally expelled.

Of Moses Maimonides it was said, "From Moses to Moses there was none like Moses." He had great influence upon the thinking of Augustine (who later became a saint) as well as Jewish thinking. He wrote *The Guide for the Perplexed*. He wrote in Arabic to reach a large audience. While traveling through North Africa he had visited a mosque and read the Koran. He was charged in an Islamic court with having been a Muslim and leaving Islam. He was on trial for his life because a sermon of Mohammed recorded in the Koran states, "If a man leaves Islam he should be killed and he will be replaced by another who is born." He defended himself by saying, "Any learned man should read the Koran," and was found not guilty. He made a living as a physician and among his patients were the rulers of Egypt.

Years ago I visited his tomb in Tiberias, Israel, high on the side of a mountain overlooking what in English is called the Sea of Galilee, which looks like a mountain lake even though it is 60 stories below sea level.

Back to the subject. Among the consequences of a clock going faster and faster, the work done in taking a charged particle along a closed path through an electric or magnetic field (or in taking a mass along a closed path through a gravitational field) is not identically zero. The longer the time of travel around the closed path, the greater the energy difference from zero. This is because of the redshift of the energy in the earlier portions traversing a longer distance.

The mass of the vacuum (so-called)

Space is not empty. Cambridge University physicist Paul Dirac (1902–1984), based on quantum mechanics, predicted the sudden appearance of electrons in space paired with positrons, which are like electrons but of positive charge. They have borrowed energy from space and quickly annihilate each other to return it. The positron was thus predicted before it was found to exist. In the meantime, with the positron-electron pair existing, the so-called vacuum has mass. The larger the universe, the greater its mass. Thus space expands itself; the larger the universe, the faster it expands. Time is created as the universe expands.

An atom smasher separates the electron from the positron (and maybe oppositely charged pairs of greater and greater mass) before it can annihilate itself. Likewise, as matter falls into a black hole it is torn apart, separating particles and antiparticles; black holes are thus atom smashers.

According to my theory of the creation of time, the stopping of expansion would cause the universe to disappear like a soap bubble bursting, but without a remnant. This is impossible.

Relationship between gravity and electromagnetism

Albert Einstein, until his last days, tried to combine theories of gravity with electromagnetism, but failed. (I add here that to add to the confusion, there are theories compatible with relativity that have physical constants like the attraction of gravity change with the age of the universe.)

Gravity attracts mass. Electromagnetism interacts with magnets and electric currents. Each in its own way distorts space-time. Thus space-time is different for gravity than for electromagnetism. Space-time is different for positively charged massive protons than for very lightweight positrons and electrons. Charge is defined by the space-time in which it exists. The problem of combining the effects of gravity with those of electromagnetism is resolved very simply by recognizing that multiple space-time can coexist.

To help you understand what I mean by coexist, imagine ghosts (which do not exist) walking through each other. If ghosts are not scientific enough for you, think of multiple image projectors, like slide projectors, projecting on any surface, transparent like glass or opaque like a painted wall. The reflected images are combined (superimposed).

Might there be a way, not yet discovered by humans, in which electromagnetism can influence gravity and be used for propulsion between solar systems? This difference in space warping and energy is a little like thermodynamics where there is a difference in temperature, enabling an electric power generator to have the energy to generate electricity.

Clockwork universe or free will?

Albert Einstein said, "God does not play dice with the universe," meaning that each event happening is fully determined by what preceded it, even if we have no way to calculate it.

Werner Heisenberg's uncertainty principle states that when you try to measure both the velocity and the position of a particle you are disturbing it so the accuracy of your measurements, on a sub-microscopic scale, is limited. This is the basis of quantum mechanics. This implies free will; that everything is not predetermined.

Einstein and a fellow physicist, Rosin, applied "common sense" to write a paper challenging Heisenberg's uncertainty principle and quantum mechanics, stating the consequences, including exceeding the velocity of light and "entanglement," of what happens at one location with what happens at another—very spooky. This is the most referenced paper on quantum mechanics. An optical corner reflector placed on the moon during the Apollo program reflected a laser beam back to the location of the laser on earth, and the "spooky" predictions of quantum mechanics were found to be true. The beam bounced from one mirror surface to the other and then toward the laser aimed at it. The laser pulse round trip was timed at the velocity of light which has been measured as 3,000 kilometers per second.

Entanglement effects exceed the velocity of light. The universe is stranger than humans could have predicted. In 1935, a physicist named Schrodinger, at the University of Dublin, thought up an experiment to illustrate what quantum theory predicts in the theory of superposition at the size level of atoms and smaller, and the apparent contradiction to what we experience in everyday life. In it, a cat is alive in a cage until the cage is opened and the cat is found dead. This is instantaneous and if the experiment is rigged so that opening of the cage is dependent upon the velocity of light, the velocity of light is exceeded; the universe is stranger than humans can imagine.

The uncertainty principle and quantum theory were confirmed and there is free will. Einstein was wrong; "God does play dice with the universe." But it is not that simple; Einstein was partially correct.

M.C. Escher (1898–1972), the great Dutch artist, created a drawing of two hands rising out of the paper holding pens drawing each other. I do not know what he was thinking as he drew this but to me it could mean God creating himself/herself in a lifeless, godless universe that had to wait for super-supernova explosions to convert a mix of hydrogen and helium to include carbon, oxygen, and other elements essential to life.

Even if only one solar system in a million in the universe we live in has evolved intelligent creatures who have developed civilizations, with the over 100 billion solar systems in the galaxy we live in, not including red dwarf stars, and more than 100 billion galaxies in the observable universe, there are more civilizations than the number of humans who have ever lived on earth.

CHAPTER 4

∞

The Nature of God – The God of Nature

For some, religion plays a central part in their lives. Observance within even the same religion may vary greatly. Over time, different belief systems have evolved. Therefore, modern day fundamentalists appear to be following rhetoric instead of searching for meaningful content. What may have once been the basic principles of a religion have, over time, been transformed into unrecognizable religious practices. Yet, different forces may influence people to strongly commit to beliefs and rituals which are far removed from the basic values and concepts of their religion. This global world is bringing people in contact with religions different from their own. In order to understand a religion, it is wise to read the scriptures upon which it is based and also read history to see how it has evolved.

An afterlife??

The idea of an afterlife was pagan as with the ancient

Egyptians who built the pyramids to house the mummified bodies of their pharaohs which were often put in boats to take them to the world to come. Their household servants were in the pyramids with them; the jobs of their servants were lifetime—the pharaoh's lifetime.

In ancient China, at the then capital of Sian, was the grave of the emperor Chin, of the regional country of Chin, who by warfare unified the country of China which is named after him. He ended the custom of burying the emperor's bodyguards with the emperor. He had individualized pottery statues of the soldiers of his bodyguard, of more than 8,000, and even of his horse and carriage, buried with him. He died in 209 BCE. This was discovered by Chinese farmers in 1974.

The pagan ancient Greeks knew that the earth was hotter as you went deeper. There were very hot springs in Greece. They described hell as a place deep within the earth where spirits went after death. The pagan hell crept into other religions. Judaism does not specifically explain an afterlife. It holds many traditions. The emphasis is on life—leading a moral life.

Over many centuries the question was asked, "Why would people behave themselves if they did not fear roasting in hell?" Those asking the question were likely revealing their own childish lack of ethics with good behavior rewarded in heaven and bad punished in hell; not good for its own sake.

The American Declaration of Independence mentions the God of Nature. The Continental Congress appointed a group to write it. The group (committee) included John Adams, later vice president under George Washington and the second president of the United States. Jefferson became the third president. Benjamin Franklin was also on the committee, and as ambassador to France from the Continental Congress, played a major role getting France to join in the war. Among Franklin's pallbearers at his funeral were his friends, the two rabbis of Philadelphia. The committee chose Thomas Jefferson from among its members to write a draft. Earlier Jefferson had taken a razor to a copy of the Bible and cut out anything he regarded as a miracle. The members of the committee were, like Jefferson, all church-going deists, meaning that reason and observation of the natural world were sufficient to prove the existence of a creator. They rejected miracles, the trinity, and revelation. The committee rejected the draft because it did not mention God, so Jefferson inserted "the God of Nature."

The language of the Old Testament has also evolved. During the Babylonian captivity, five centuries BCE, following the Babylonian destruction of Solomon's temple, the Jewish leadership in Babylon wrote the oral tradition and translated both it and the written scripture of Moses, including the ten commandments, into Babylonian script which became today's Hebrew. They also declared that with the destruction of Solomon's temple, every man had become his own priest and was to wear a head covering that was the symbol of the priesthood; thus Leonardo da Vinci's and other European paintings of the last supper, which was a Passover observance (Seder) were inaccurate because they showed bare heads, as well as European faces, not Middle Eastern.

The translation of the Bible from language to language distances us from original meanings. Names in Hebrew are not just names. They have meaning. For example, near the very beginning of the Bible, God says, "Let us make man in our image." This is the basis for the name "Michael." His name is interpreted to mean "Who is like God?" The feminine variation is Michelle. The name "Adam" means both "red" and "earth." The imagery of man is that he was created of red earth. The name Benyamin means "son of my right hand." In the Greek translation, the y is replaced by j.

Yoshua (*Jesus*, in Greek) was born during one of four Jewish religious observances specified earlier in the Bible during which Jews were to bring sacrifices to the tabernacle, later the temple. None were in the winter around Christmas. Centuries after the crucifixion, the Catholic Church in Rome decided the birth of the infant Jesus should be celebrated. They placed the brit (circumcision) of Jesus on the Roman new year's day so the Romans could continue celebrating, but they were told they were celebrating the brit of Jesus. Counting back seven days from the brit, placed the birth of Yoshua on Christmas Eve.

The favorite disciple of Jesus was Simon (somehow called Peter in Greek translation.) Peter was brought to Rome by the pagan Romans, tried on the charge that he practiced the Jewish religion and convicted. He was sentenced to be crucified. To be different from Jesus, Peter asked his executioners to crucify him upside down and they did. Thus Peter was the representative of Jesus in Rome. The bishop of Rome's claim to be head of the Catholic Church (pope) is based on Peter having been brought to Rome to be tried and crucified.

The inventor of anti-Semitism was the pagan Roman emperor, Constantine I, who observed that the pagan people of the Roman empire only paid lip service to their gods but those who like his mother, Helena, who had become Christians were enthusiasts. He agreed to make Christianity a legal religion of the Roman empire provided the Roman culprits who crucified Jesus were not blamed for it. Christians who blamed Jews for not accepting Jesus as the messiah (*Christ* in Greek) got the blame shifted to the Jews.

In time scale Constantine I and Constantine II were about two centuries before Mohammed and jihad. Constantine's mother, Helena, had become a Christian. She went to Jerusalem, and walking the streets, declared what were the stations of the cross. The Church of the Holy Sepulcher was built on the site she had declared was where the body of Jesus lay temporarily before going to heaven. She also went to what was thought to be Mount Sinai and declared which bush was the burning bush where Moses heard the voice of God.

At the end of his life, Constantine I, feeling he had nothing to lose, became a Christian. Constantine's successor was his son, Constantine II, who also at the end of his life became a Christian.

Stress shortens life span. Sincere prayer reduces stress. Is mechanical prayer, simply mouthing words, useless?

The god of nature has great patience. Before life could exist anywhere in the universe billions of years had to pass during which large stars consumed their hydrogen, turning it into more helium, and then became supernova exploding and creating oxygen, iron, carbon, calcium, magnesium, sodium and all the other types of atoms necessary for the building blocks of life.

The most quoted book is the Bible. To be an educated person, even if you are an atheist or agnostic, it is important to read and study the Bible, both the Old and New Testaments to understand quotes and their contexts.

In Egypt, at the time of the exodus, the Jewish community believed in God who acts through nature. They believed that God guided nature in creating the ten plagues that Moses attributed to God. Pharaoh was skeptical but after the tenth plague, the deaths of the first born, he relented and let the Jews go.

The plague of darkness was likely the result of a major detonation of the volcano Santorini in the southern part of the Aegean Sea near the coast of Crete that demolished the Minoan civilization, burying it deeply under volcanic ash and creating a tsunami that destroyed their ships and coastal cities. This volcano eruption was dated by radioactivity to have been in 1613 BCE plus or minus seven years. This is in rough agreement with the statement that the exodus began 430 years after Jacob arrived on the selfsame day.

Moses led the exodus on the full moon when the tides of the Indian Ocean were highest. The wind direction changed and Moses changed his path, probably southward, to go by the Sea of Reeds. As the Jews and those accompanying them reached the Sea of Reeds, the wind combined with the tide drove the water out and they could walk across on land. Pharaoh had changed his mind about letting them leave and had sent horse-drawn chariots after them. In what is called a seiche, when the wind stopped blowing, the water rebounded back, drowning the Egyptians in their chariots. Moses timed the crossing so that when the Egyptians following them were in the crossing, the water would start quickly coming back, bogging chariot wheels in wet earth. When Napoleon was in Egypt, he was in the Sea of Reeds and had to rush out before the tide came in to avoid being drowned. This phenomenon of very high tides rushing back in exists today at the Bay of Fundy between Nova Scotia and New Brunswick in Canada. Since the Suez Canal was dug, it no longer exists in Egypt. The Sea of Reeds no longer has extreme tides.

In the summer of 1954, we were living in an apartment in Chicago overlooking Lake Michigan beyond a park. We were not looking out at the time, but there were people fishing, sitting on concrete piers protruding into Lake Michigan perhaps six feet above the water. A strong wind from the south stopped and the sloping water from the about 400-mile length of the lake rushed back, drowning some of them—an act of nature.

As a young man, Moses had observed an Egyptian severely beating a Jewish slave. He killed and buried him. A little later Moses saw two Jews fighting. When he tried to get them to stop fighting each other they asked him if he would try to do to them what he had done to the Egyptian. So Moses fled into the desert and waited until he heard his enemies had died, then returned to Egypt and confronted

Pharaoh. During the exodus he brought down the tablets with the ten commandments from Mount Sinai.

There is no commandment, "Thou shall not kill." That is a mistranslation or over time a change in the meaning of the word "kill." The commandment is, "Thou shall not murder." Some, including pacifists, who do not understand this become conscientious objectors, and even do not defend themselves if attacked, letting themselves be killed, thus allowing their attackers to continue on and kill others.

The book of Numbers is largely a census of men over 20 able to wage war. Over 20, to give them a chance to have a child to represent them in case they were killed in combat.

Moses was not a dictator. As described in the Bible, he was overruled by the elders of the twelve tribes on the laws of inheritance.

Human sacrifices by pagans were common in the time of Moses. I speculate that the reason Moses had his brother, Aaron, as high priest, perform animal sacrifices, was to avoid human sacrifices. At the temple in Jerusalem, on the day of atonement, lots were drawn on two goats. One goat, the scapegoat, was sent into the wilderness to carry the sins of the people away, the other was sacrificed.

Aramaic was the language of Abraham before he moved to Canaan. Hebrew was the regional language of Canaan, Phoenicia and its colonies like Carthage. The original language of the people of the Mycenaean civilization in Crete, before their conquest by the Greeks invading from the north, was a dialect of Hebrew. Jewish marriage contracts, called *ketubas*, are written in Aramaic which was the language of Babylon. The book of Daniel in the Old Testament is the only surviving book of the Bible written in Aramaic. The book of Esther (meaning *hidden* in Aramaic) was written in Hebrew probably by her uncle Mordechai even though the spoken language of the Jews then in exile in Persia was Aramaic.

The original writings of the disciples of Jesus were destroyed during a rebellion against Roman rule and only the Greek translations survived. An example of a miracle of translation is "walking on the water." The same Hebrew expression meaning "on" also means "next to." Probably likewise for Aramaic which is a closely related language. In synagogues in Greece where the Old Testament was studied, it was noted that the prophet Isaiah (about three centuries after King David and seven centuries before Jesus) had prophesied that the messiah would be a descendent of King David and the son of a young woman. The Romans made a fetish of virgins. When in Rome a woman was sentenced to death, a duty of the jailers was to make sure that she was no longer a virgin when she was executed.

The New Testament states that the descent of Jesus from King David is traced through his father Joseph, not his mother. His mother becoming a virgin was a miracle of translation from Greek. During his crucifixion, according to the New Testament, his executioners ridiculed him, placing a crown of thorns upon his head and calling him "King of the Jews."

On the cross on Friday afternoon Jesus cried out a quotation from the Bible, Psalm 22, "My God, my God, why have you forsaken me?"

The country with the largest Muslim population now is Indonesia. The reason is the eruption of the Krakatoa volcano in 1883 with the energy of an estimated 200 million tons of TNT equivalent. Muslim missionaries told the people of what was then the Dutch East Indies that their pagan gods did not protect them and they needed the protection of Allah.

Islam is based on the Koran which is a collection of undated surviving sermons of Mohammed recorded by his relatives. The sequence is with the longest first to the shortest last. The Sunnis and Shiites often dispute over who is the rightful successor of Mohammed, spilling blood.

Sometimes they cooperate against those who are non-Muslim. Mohammed stated that martyrs who die for Islam are rewarded in paradise, which he describes in glowing terms. This motivates fundamentalists. The Koran should be read in order to understand Mohammed's words which include violent doctrines against non-Muslims. Nonbelievers should be killed, therefore, conversions can be forced. For those who leave Islam, the appropriate punishment is death. Legally binding agreements with non-Muslims do not have to be honored. Mohammed spoke of his making a treaty used to deceive the enemy. These violent concepts found in the Koran express the harsh side of the religion.

Buddhism teaches meditation which reduces stress. You can be of any religion and even an atheist and also a Buddhist. A Muslim is very unlikely to also be a Buddhist. Buddhism has evolved rituals like a religion and is therefore often mistaken for a religion. There is no god in Buddhism. The original Buddha came from a prominent family in what is now Nepal, north of India about twenty-five centuries ago. He left his family and went out into the world to meditate. To keep from starving he became a beggar. Buddhism accepts the idea of reincarnation; being reborn as an insect, then as an animal, then as a human. One definition of *Buddhi* means "to wake up, to understand." The Chinese years are named after the animals believed to have been with him as he died, like the year of the rat, goat, monkey, rooster, dog, pig. These were supposedly the creatures who were with Buddha as he was on his deathbed, dying peacefully at the age of 80. The years repeat in cycles. The Chinese new year is celebrated at the second new moon after the winter solstice and thus falls between January 21 and February 19. The Chinese simplify birthday celebrations by everyone celebrating on the Chinese new year's day.

When the previous Dali Lama died, a search was made to find the male baby in a Tibetan Buddhist family who was born closest to the time after the previous Dali Lama died. That baby was brought up to become the new Dali Lama. The Dali Lama is the leader of Buddhism. He is in exile from Tibet, living mostly in India. He visited Israel to discuss how Jews had managed to survive as a distinct people living for two thousand years mostly in a diaspora—the problem facing Tibetan Buddhists. He created the Mind and Life Institute which uses neuroscience to study the effects of meditation on the brain itself. It was found that the regions of the brain most active in meditation enlarge with many long periods of meditation. Today biofeedback in which electrodes are placed on the scalp displaying brain waves to the person who is meditating. The person can then meditate in a way to control his/her brain waves, making them less active. In modern society, a person trying to emulate the Buddha—abandoning his wife and children and his wealthy parents, living largely outdoors, meditating, and begging so as not to starve, would not be highly regarded by many.

Pagans

Pagans are well-represented in modern culture. For example, the days of the week are named after pagan gods, except Friday which is celebrating freedom for women as with the female name Frieda. Wednesday is named for the Germanic god, Woden, who lived in trees. There is an old saying to use when you want good-luck: "knock on wood."

The names of the days of the week are named after Roman or Germanic gods.

Names of the days

    Sunday – Sun's day
    Monday – Moon's day
    Tuesday – Tiu's day

Wednesday – Woden's day
Thursday – Thor's day
Friday – Freya's day
Saturday – Saturn's day

In the winter, as the days became shorter, Germanic tribes living north of Rome had a human sacrifice to make sure the days would not keep getting shorter and shorter and the sun did not totally disappear. It always worked and they were satisfied. They resented the Vatican sending missionaries to tell them they did not have to continue having human sacrifices because Jesus had sacrificed himself for them and that as Christians they could continue decorating trees but now it was in honor of the birth of Christ.

The Puritans in England, who believed that Christianity must be pure in its observance, revolted under the leadership of Oliver Cromwell and chopped off the head of the British king. They enforced strict observance of no Sunday sports—like rolling balls on grass. The British threw out the Puritans and selected a new king. A town in Connecticut was named after Cromwell.

While they ruled Boston and elsewhere, the Puritans banned the celebration of Christmas as Pagan. In the winter they rounded up the Indians to get rid of them and placed them on an island in the Boston harbor to freeze and starve to death.

This was not as bad as what the Spanish settlers had done on Hispaniola (the island Columbus landed on and which is now divided between Haiti and the Dominican Republic). Columbus had been appointed governor of the island as a reward for his having claimed it for Spain. Many of the natives of the island did not survive the diseases which the Spanish had brought with them. Their immune systems had not evolved to survive these diseases. Although they had welcomed Columbus, they resisted becoming slaves of the

Spanish settlers. The Spanish rounded up the natives and tied them to stakes in groups of 12, corresponding to the 12 disciples of Christ. They then burnt them at the stakes. If they knew of this, the disciples would have rolled over in their graves. The natives were replaced by disoriented, obedient slaves from Africa.

Columbus objected and was imprisoned. When Cortez led the conquest of the Aztec empire in what is now Mexico and brought back much gold, Columbus was released and returned to Spain with honors.

Pagan celebrations today

Halloween and Mardi Gras are both of Pagan origin, as is the custom of decorating Christmas trees—in honor of the Pagan god, Woden, who dwelt in the trees. Halloween is celebrated the evening before All Saints day. Mardi Gras began as a Pagan Roman fertility celebration. The early Christians, in order to convert Pagans, tried to turn these Pagan events into Christian festivities continuing with masks and costumes.

Inherited memories

A newborn infant has the instinct, remembers, to suckle on its mother's breasts. I have heard stories of individuals who thought they knew the layout and appearance of the home their parents had lived in before they were born. I cannot automatically dismiss this as untrue. Much of our DNA heredity was not understood and was dismissed as "junk" DNA. It is now better understood with much of it being inherited behavior, instinct.

After your life do you have contact with others, family, friends? Do you have thoughts, dreams? Do you sleep? All the time? Or is it oblivion? "Rest in peace."

CHAPTER 5

∞

UFOs

Some will probably ridicule me for including this subject, but keep an open mind. A June 1947 report by pilot, Kenneth Arnold, flying his small plane near Mount Rainier over the state of Washington, described seeing nine disk or saucer shaped objects. He later added that one of the objects was crescent shaped. There were over 800 reports by Americans in the summer of 1947 of unidentified flying objects (UFOs). Of course some of these reported sightings were misinterpretations of what they saw, but not all. In December 1947 General Curtis LeMay requested an update on flying saucer sightings. The military was concerned about numerous close encounters of our aircraft with UFOs. The information was classified because the military did not want the public to be concerned when they could not do anything about it. Also, the federal government providing mail addresses to report sightings indicates they took the issues seriously. In addition, deliberate misinformation was provided to keep the situation confused.

In July 1947 there were reports of a UFO crash near Roswell, New Mexico. I initially regarded these as garbage because of the commercialization of these events in Roswell for tourist dollars. But I changed my mind when I investigated further. The Roswell military base public information officer issued a report, published in the local Roswell newspaper, about a crash of metal disks in the Roswell area. An official denial the next day, regarded by many as a cover-up, stated what was recovered was actually debris from a special weather balloon.

As I interpret the reports, necessarily a speculation because of the deliberate misinformation, I give credibility to the local funeral director. He stated that he delivered child sized bags to the local airbase. He further reported that his friend, a nurse, told him of assisting two physicians, flown in from San Diego, in performing an autopsy on a body of a creature that was not human. It had anatomy unlike any creature on earth, had been chewed by animals, and stunk very badly from days of decay. The nurse told the funeral director she had been warned not to speak to anybody about this or she would face serious consequences.

Within a few days, she disappeared and records that she had existed also disappeared. There were many reports at other times and places of such warnings. A man on his deathbed told a story that replaced what he had said many years earlier had happened. To add to the confusion, nearly 20 years later a film was made of a reconstruction of this alien autopsy, passed off as the real thing, and then admitted to be a hoax. I believe the original report by the funeral director was credible.

What I think really happened was that in a thunderstorm, one so-called flying saucer was struck by lightning and crashed into another, both then hitting the ground. In one report there were survivors who were put into a "bubble" building in California to protect them from earth germs and were later visited by President Eisenhower.

Barry Goldwater was the presidential candidate; a senator from Arizona, and combat pilot in World War II. After the war, he was a major general in the air force reserve. I met him at a meeting of an industrial group founded to support the air force. Years later I saw him on television describing a meeting in which he had asked his good friend, Curtis LeMay, air force chief of staff, for permission to visit the building at Wright-Patterson Air Force Base in Dayton, Ohio, that housed the "Foreign Technology Division" and where it was rumored that the bodies of space aliens were stored refrigerated and under glass. Curtis LeMay screamed at him, "Don't ever ask that, even I cannot do that." Years later I spoke to a son of Barry Goldwater. He was a believer in space aliens and described large flying objects over Phoenix at night which the air force had dismissed as flares dropped during practice maneuvers.

President Truman had expressed concern about public reaction to a possible landing of a "flying saucer" on the grass of the Washington Mall or the White House lawn. Before World War II, on Halloween evening, Orson Wells, on the radio, played a science fiction story written by H.G. Wells about Martians landing in New Jersey. Every 15 minutes he interrupted to say it was a Halloween special but some people in New Jersey panicked, loaded their cars and fled. We were in New Jersey listening to another radio program that night, ventriloquist Edgar Bergen with his wisecracking dummy character, Charlie McCarthy, which we could not see but pictured as sitting on his knee. President Truman was reported to have an advisory group called the Majestic Twelve that recommended that UFOs be neither confirmed or denied, that misinformation be leaked, and that there be studies to show the public something was being done. An anonymous mailing that described the Majestic Twelve, labeled top secret, was analyzed as to typewriters of the period to determine if it was a forgery but nothing was found wrong.

In the USSR, a group of scientists including an astronomer was appointed to investigate the existence of UFOs.

More recently the Vatican appointed an astronomer to avoid a repeat of its Galileo fiasco. The Vatican astronomer was asked to form a group to advise the pope what to do if there is communication from space aliens. The answer to that is simple. Some place in the New Testament, Jesus is quoted as saying, "My father's house has many rooms," but there might be objections to this as undermining the authority of the Vatican.

After dinner on a Friday early evening in the second half of May 1952, after the sun had set but the sky was still light, my wife and I and my visiting sister went for a walk on top of a hill in the Congress Heights section of Washington, DC. Looking south toward the Potomac River we saw a bright green light. The shade of green was that of the copper flame test in a chemistry qualitative analysis lab, which is the same color as in fireworks. It was moving in the apparent direction of the flight path taken by aircraft landing at Bolling Field, an air force base.

Bolling Field was later closed after a plane being flown by a Bolivian pilot in training, under military air traffic control, crashed into an airliner near Washington National Airport (later renamed after President Reagan) under Federal Aviation Agency (FAA) control—two different agencies, the same air space, each agency defending its own turf. Stupidity. This situation was finally ended by that crash.

I could not judge the distance to the light. There was silence and there was only one light, not the green and red much dimmer lights on the wing tips of aircraft and the brighter white lights illuminating forward. I wrote this in a letter I mailed to a post office box the government had established in Virginia for reporting UFOs. (The PO box shows that in spite of denials, the government was taking UFOs seriously.)

There were reports that airline pilots flying south of Washington over Virginia were seeing UFOs but were forbidden by the airlines from calling them to the attention of passengers because that might frighten them and reduce air travel.

In July 1952 there were a number of unidentified objects detected by Washington National Airport radar. The FAA dismissed these as ground reflections of the radar beam caused by unusual weather conditions bending the radar beam down, as though the experienced radar operators would not know the difference. In this same time period, in Washington in broad daylight, an initially skeptical man was convinced of the existence of UFOs when he saw a saucer shaped metallic object against a clear blue sky. This was all well-reported in the *Washington Post*.

I am aware of a number of UFO sightings over the years by very credible witnesses, aircraft crews, and engineers. Radar has tracked UFOs coming up out of the sea and going into earth orbit around the earth. These were not in the press.

Talking to a pilot who had retired from the navy, he told me he had been on an aircraft carrier in the Caribbean. In practice maneuvers two squadrons had flown out in opposite directions. Upon returning to the aircraft carrier the crews of each squadron reported that metallic disk shaped objects had circled around them.

In the 1970s, a credible man I knew well who lived in Pound Ridge, New York, told me that a boy who was dependably truthful told him he had seen a flying saucer shaped metallic object floating in a pond in Pound Ridge. There were two "men" in space suits holding a hose into the water apparently sucking up water into their craft.

In 1987, I was at my 40th official college reunion and speaking to an engineer from General Electric, Utica, New York. He told me there had been many colored lights in the sky in that region.

There were other observations by my own family and friends. In the spring of 1977 my middle daughter and two of her friends were studying at Syracuse University in upstate New York. They were out for a drive in the countryside on their way to go camping. Far from any buildings, in a remote location, they noticed an extremely bright white light in the distance at the top of a small mountain. Since they were "in the middle of nowhere" they all started guessing what the light belonged to. They came up with the idea that there must be a school with an outdoor field lit up for some sports event. They continued up the mountain road. When they arrived they were surprised to see that there was no school or any sign of civilization at all. They looked to the right of the road and saw a flying saucer. My daughter was the driver and because she was curious she stopped the car to observe. The three had a clear view of the saucer hovering over a field. Its bright white light was combing the field. My daughter attempted to get out of the car but stopped because the other two passengers were grabbing her and hysterically beckoning her to drive as fast as possible away from this UFO. She started to drive away. It then combed over their vehicle with its lights and silently flew away. They were shocked that it looked exactly like those described in literature and seen in films. Many years later one of the passengers of that van retold this event to me, confirming what my daughter had previously told me.

While visiting our home in her late 70s, my mother was telling our family stories of her childhood. We got around to the subject of UFOs. She was living in the countryside just outside of Smithville, Pennsylvania, south of Scranton where her father was operating a bar selling whisky to the local coal miners. She told us she saw something that looked like an orange football low in the sky. I asked her if it could be the setting sun and she said no and was upset at being questioned. She said her next door neighbor who was a deputy sheriff, stepped outside with his gun and shot at it, but nothing happened. My mother was born in 1893. The incident she was describing may have been in 1898 when there had been many strange sightings in the sky over the United States.

On July 11, 1991, in the Mexico City area there was an eclipse of the sun. Records show 17 people who did not know each other had video cameras out to take videos of the eclipse. I remember one broadcast on TV at the time that showed a metal disk between the camera and a tall building in Mexico City. Also in the 1990s there were many more UFO incidents in Mexico. A number of airline pilots reported UFOs below them, sometimes requiring them to make evasive maneuvers. In one case the landing gear was damaged in a collision with a UFO below the plane.

Long before recent centuries there were many sightings. About 1500 BCE, engraved in hieroglyphics, there was a report of a pharaoh commenting that circles of fire in the sky are more common now. The ancient Romans reported flying shields.

In the Bible, Lot, a nephew of Abraham, was living in Sodom which archeologists now place at the southern shore of the Dead Sea. Climate was different then, more rain, because the earth was still in the process of coming out of the last Ice Age and farming, in what is now dry desert, supported life. My interpretation of what took place in Sodom is that "angels" (space aliens?) were using laser technology as weapons. Genesis 19.11, "And they smote the men that were at the door of the house with blindness, both small and great; so they wearied themselves to find the door."

In the Bible, the vehicle that lifted and returned Ezekiel (Ezekiel 1.16) about five centuries BCE, was described with a limited vocabulary in terms of features of animals. It was described as having wheels that did not turn as it steered itself. To understand this, think of a string of figs rolled up into the shape of a wheel; the "figs" turning to move the vehicle sideways while the wheel turns for the forward component of motion. An engineer who had worked on the Apollo program applied for a patent based on the description of the wheel.

A century after King David, nine centuries BCE, in what may be a fourth hand description (Bible 2–Kings) the Prophet Eliyahu, (in Greek translation, *Elijah*) is described as taken up into the sky in a chariot of fire pulled by horses of fire. This is the inspiration for the song "swing low sweet chariot, coming to carry me home". At a *brit* (circumcision ceremony) a chair is reserved for Eliyahu to witness. At a Passover Seder a cup of wine is filled for Eliyahu and during the ceremony a door to the outside is open to let the spirit of Eliyahu enter. In both Jewish and Christian tradition, Eliyahu/Elijah is to return before the coming of the messiah (in Christianity he is Jesus).

In reports of space aliens, two different body types are described. One in the United States is child-size with no thumbs, ridges on four fingers to help hold tools, grey skin, and black eyes with no pupils. My informed speculation is that this body type has copper based blood rather than iron based. It has compound eyes absorbing light (like insects on earth), with no pupils, and requires much more brain capacity to form images.

The other body type reported from Russia and Argentina is about two meters tall (over six feet). The report from Argentina mentioned hard-shelled, claw-like hands. What we can know from evolution on earth about the evolution of "people" in civilizations elsewhere in the universe is discussed in part of Chapter Ten – Evolution.

Space aliens, in many variations, are very unlikely to have evolved the same hearing and vocalization as humans. If listening to human music, some may prefer drum beats and clicking sounds, others jazz, others classical, still others singing. Some may be unable to speak loudly and have only "a still small voice" (Bible 1 Kings 19).

Eyewitness evidence versus probability

The explosion and crash of TWA flight 800 taking off from JFK Airport in New York on July 17, 1996, at 8:31 pm is an example. I looked up the date and time. The rest of what I am now writing about is from memory.

There were two credible witnesses who saw a shooting star (a white hot, tiny rock about the size of a pea) rapidly coming down and when it reached the aircraft it exploded. That means it not only hit the aircraft but a partially empty fuel tank with a mixture of air and fuel vapor in the tank that was ignited by the rock, (white hot from friction with the earth's atmosphere). One witness was the pilot of a C-130 cargo plane of the Connecticut Air National Guard on a training mission flying over Long Island. The other was an engineer walking his dog on a beach who reported the same thing. About two days after the crash I read their reported stories in the *New York Times* which also reported many garbage stories about missiles going upward. The National Transportation Safety Board (NTSB) and FBI both investigated. Most of the wreckage was recovered from shallow water off the coast of Long Island and reassembled in a Grumman Aircraft facility.

Long before I retired in 1990 I had made many business trips to Grumman. I was among many technical people writing for the NTSB and was informed that the FBI was telling them to regard the eyewitness reports of the shooting star causing the explosion as credible. I wrote, "If when driving on the interstate highway system you pull into a rest stop and look at a car parked next to you, what is the probability that a car of that make, model, color and license plate is there? It is 100% because it is there." The

NTSB hired a man whom it considered a statistics expert who issued a report that a shooting star as the cause of the explosion was so unlikely that the eyewitness reports should be disregarded.

Reports of UFO sightings, some recorded by cameras with tall buildings behind them, some with the image of the UFO being recorded on only a single frame, should be treated as credible as well as the crash of TWA flight 800 having really happened.

What could be the mission assigned to space aliens before they were sent here from relatively nearby (a few light-years) solar systems in our galaxy? I speculate that their mission recognizes the danger to them of exposure to earth germs to which we have evolved immune resistance.

Humans tend to be arrogant, feeling that if we don't know how to accomplish something, no other civilization in the universe knows how. For example, two things we don't understand are the speed of travel and avoiding ice dust. What would be the means of propulsion to be able to travel at half the speed of light without killing the occupants of the vehicle? Collisions with space dust would destroy the vehicles if not protected by a space shield (not yet designed by our technologists).

But the evidence is that somehow space aliens are here and have been for many of their generations, which could have evolved to be many times ours. Evolutionary pressure on humans to reproduce before we die from disease or violence and our evolution including the symbiosis of mitochondria with other life are all pressures that shorten life on earth.

I speculate that their mission is to wait, study us, and when we are starting to kill ourselves off, as with nuclear weapons, intervene and prevent us from becoming a danger to their own civilizations.

Looking at the earth from space, the most likely region to contact and influence people is where Eurasia and Africa come together. Given human religions and history it is the Temple Mount in Jerusalem, not the grass mall or White House lawn in Washington, DC, where they will land. Since we on earth are likely being monitored, they know that we believe that the messiah will come to the Temple Mount in Jerusalem. If they land on the Temple Mount, this may be a self-fulfilling prophesy.

CHAPTER 6

∞

The Air We Breathe – Green Mythology as Taught

The real source of the oxygen

Just as you may have been taught Greek mythology as a child, you probably were also taught "green mythology," i.e., most of the oxygen in our air comes from vegetation. The "beautiful balance" in nature between man and plants is critical. We depend on the oxygen that vegetation releases. This idea is attractive, thoroughly entrenched, but not true. My writing on green mythology will be controversial. So where does most of the world's oxygen come from?

As plants grow, they absorb carbon dioxide from the air and release oxygen. But they still have carbon until they decay. Then they are absorbing oxygen from the air. This is a balance in nature.

Long ago, I dumped some small fallen branches in the woods behind our home. Twenty years later I picked them up. Their bark was still intact but largely empty, hollowed out by slow oxidation (cool burning) turning the wood back into carbon dioxide. Carbon taken from carbon dioxide in the air by photosynthesis to help form plants is returned to the air by oxidation. The air is about 20% oxygen.

Part of the green mythology includes the idea that as modern man has built, he has caused a significant imbalance between carbon dioxide and oxygen. The destruction of plants has caused a significant increase in the carbon dioxide of the air we breathe. Therefore, to be a responsible being, you need to care about nature. In actuality, we can and have accurately measured the increase of carbon dioxide. It has gone up but only slightly. By taking samples of ice from deep within glaciers the precise amount of carbon dioxide trapped in air bubbles can be measured. The ice acts as a time machine. The deeper we drill, the further back in time we can analyze the amount of carbon that was once in the earth's air. This research has been done in different locations such as Greenland and Antarctica to cross compare. These ice samples are such important historic guides that they are stored in America. The change in the level of carbon dioxide has been measured by parts per million. That is how insignificant it has been. In 1750 just before the Industrial Revolution we would expect a significantly lower level of carbon dioxide than that of today. This is not the case. Then, it was 280 parts per million. In 1950 it was about 310 parts per million. Now it is about 401 parts per million which is still just a tiny fraction of 1% of all the air.

There is an extreme shortage of carbon dioxide in our atmosphere to account for the one-fifth oxygen claimed to have been produced by green plants. The numbers simply don't add up. Most of the oxygen that has been formed over the centuries is not in the air. It is now part of the rocks and the soil. Look at the red rocks in the walls of the Grand Canyon and other red rocks around the world. I do not have the numbers but obviously there is far more oxygen bound up in red rocks and in soil around the world than exists in our atmosphere. With the weight of the earth's atmosphere being only that of 30 feet of water; the atmosphere is only about one-fifth oxygen or roughly that of six feet of water.

I became aware that in the late 1800s, as chemistry was beginning to be understood, some people explained the source of the oxygen in the air we breathe. Some water vapor reaches the very upper levels of our atmosphere. Gamma ray (extreme x-ray) radiation splits the water molecules (two hydrogen atoms combined with one oxygen atom) into hydrogen and oxygen. A hydrogen atom has a unit weight of one compared with an oxygen atom's unit weight of 16. In a gas mixture there is an equipartition of energy among gas particles, meaning that on the average a hydrogen atom has the same energy as an oxygen atom. If you crash a car into a telephone pole at 80 miles per hour you are crashing it with four times the energy as at 40 miles per hour (energy is proportional to velocity squared, not doubled). Thus a hydrogen atom in a gas mixture has four times the velocity of an oxygen atom on the average and hydrogen atoms, some moving faster than others, gradually escape the earth's gravity into outer space leaving oxygen atoms behind.

The earth is full of the wonders of nature. They should be respected not because of our dependency upon them to serve us. We should appreciate the wonders of nature in their own right.

CHAPTER 7

∞

Climate Squeal

Man's role may not be causing only warming. It may also in some parts of the world be causing cooling. Climate is always changing. The earth's orbit is an ellipse—a stretched out circle. The earth is like a wobbling top. The wobble is caused by the earth's attraction to the other planets, predominantly Jupiter. The wobble is greatly reduced by the earth being part of the earth-moon system which keeps the axis of the earth, the poles, from wandering around the surface of the earth faster than vegetation can adjust to climate change.

Climate change can be useful. A reduction in rainfall was the precondition that enabled man to start walking. A drying out of forests in East Africa about two million years ago killed off many trees, forcing our ancestors, who largely lived in trees for safety, to walk between clumps of trees. A volcanic eruption covered a footprint in mud under lava, enabling radioactive dating. It was shaped like a modern footprint, but smaller than that of modern man. We are walking because of climate change.

Climate change largely begins as snow gradually accumulates from year to year at higher elevations around the Arctic Ocean reflecting the sun's energy back into space. With this reflection of light, less energy is being absorbed. Evaporation from an Arctic Ocean that is partially ice-free part of the year speeds this process. This evaporation causes more snowfall on higher elevations around the world. This heavier snow takes longer to melt. Eventually the glaciers will stop melting. Over time the accumulation becomes deeper and deeper. For example, New York had been repeatedly under glaciers which were about two miles thick.

In this century, between the years 2001 and 2200, the earth is closest to the sun in January and farthest in July. We are presently in a period between Ice Ages. The coming Ice Age will gradually begin as the wobble shifts the time of the year the earth is farthest from the sun to coincide with the time of the year the northern hemisphere is tilted to have the shortest days.

As the earth wobbles back, and receives more solar heating, the glaciers begin to melt. The time between Ice Ages is shorter than the time when the earth is covered by ice.

We have a shortage of carbon dioxide in the air because rain and snow have absorbed it. As rain and snow fell, acid rain reacted with the rocks of the Himalayan Mountains. The Ice Ages began after the Himalayas rose.

When you are swimming outdoors on a bright warm day and step out of the water you freeze as the water evaporates off your skin. Similarly, you can freeze from global warming.

To understand "positive feedback" consider an audio system where sound from a loudspeaker reaches a microphone and we hear a loud squeal—runaway positive feedback. The speaker places his hand over the microphone while someone adjusts the amplifier.

There is more carbon locked in the Arctic than is now in the atmosphere. Long accumulated decaying bacteria, which is now thawing from the tundra grass is producing methane gas which bubbles up from the ground. About half this methane is oxidized into carbon dioxide in about each seven years; thus seven years is called the half-life of methane in the earth's atmosphere. Arctic warming is twice as fast as in the temperate zone. The more warming, the more bubbling; this is called positive feedback. With runaway positive feedback there can be no "hand-over-the-microphone" to stop runaway climate squeal, not even if all fossil fuel use suddenly stopped, which obviously will not happen. Dire consequences—population plummeting.

Snowball earth

Imagine the earth looking like a giant snowball. There has been snowball earth in the past. About 850 to 635 million years ago the sun was much dimmer, land was under thick ice and oceans froze over deeply. Life survived near warm water vents in the ocean floor such as along the mid-Atlantic ridge. Carbon dioxide from volcanoes, bubbling up, eventually broke through the ice crust. Life evolved. The danger now is that with the Arctic thawing, more of the year's evaporation will fall out as snow on higher altitudes in a ring around the Arctic Ocean. The snow will reflect more sunlight and some will accumulate from year to year bringing the coming Ice Age thousands of years earlier. Glaciers will extend farther south than in the previous Ice Age, and so on, with shorter interglacial intervals. During the most recent glaciation, Central America had a temperate climate, as evidenced by pollen typical of the United States mid-Atlantic region today. After a number of Ice Age cycles, the climate in Central America will again become similar to that of present day New England. Over a number of Ice Ages, the extent of the glaciers will grow toward the equator.

Some scientists find evidence that the present series of Ice Ages resulted from a collision between an island in the Indian Ocean which drifted northward on the earth's crust and collided with the mainland of Asia. This process is called plate tectonics. Similarly, over the past 66 million years, the North Atlantic has widened from about one-tenth its present width and the Pacific has narrowed, crumpling the shoreline, raising the mountains along the backbone of North and South America. The evidence for this includes the sea shells and other remnants of the sea bottom found on the tops of the Himalaya Mountains. Also, fish in the Amazon River are closely related to fish swimming in the Pacific Ocean showing that before the South America land mass rode out over the sea bottom the Amazon flowed into the Pacific.

Carbon dioxide in the earth's atmosphere was greatly reduced. It was absorbed in rainwater and formed acid rain (like soda pop) which fell on bare rock reacting with it, flowing downstream out to sea. The greenhouse gas, carbon dioxide, was thus depleted causing the climate to become colder. Greenhouse gas lets in radiant energy from the sun and keeps it from being reflected out. Supporting evidence that the Arctic was once much warmer are the fossils of skeletons of crocodile/alligator-like creatures found on the shores of the Arctic Ocean. We are now remedying this carbon dioxide depletion by burning fuel. People are complaining about too much carbon dioxide and other greenhouse gases in today's atmosphere. However, carbon dioxide had been greatly depleted in the past. We are only partially replacing it by burning fossil fuels. This probably delays the coming of the next Ice Age.

But on the other hand—politicians dislike scientists stating there are uncertainties in their advice—most of the Arctic Ocean ice that survives the summer melting has been measured as much thinner and much is already gone. As there becomes more exposed sea surface, evaporation will increase. More clouds will both reflect more sunlight back into space and have a greenhouse effect reflecting heat radiation back toward the earth at all times year round. This might even be delaying the coming Ice Age. And the portion of the Antarctic ice shelf floating over the ocean has been measured as growing thicker as water below freezes on to it.

Jet stream

First a digression. In 1968 I worked for an electronics company that was acquired by another much larger company. I was assigned to help them write a proposal to work on a United States government program for weather forecasting that would serve the needs of all federal agencies. We won, and I was directed to manage the work of our division part time, which turned out to be about one day a week for about ten years. I learned a lot. In those days the meteorologists (weathermen) made short-term forecasts by observing from which direction the wind was blowing and the measured wind speed. Then, phoning one or more weather stations (usually at an airport upwind) and asking, "What is the weather there?"

A computer we were building for the program was specified to use transistors instead of vacuum tubes. Two-transistor chips had become available—2,200 chips were required. The various government agencies required the same information but in different formats (sequences). We could simplify the computer if it was built with only a single format. But the lowest ranking federal official who could agree to this was the president of the United States, so we gave up on having only one format.

To help keep myself informed, I was reading the *Journal of Geophysics* in the company library. There was a very interesting errata (correction). Weather data was stored on International Business Machines (IBM) punched cards stored at Suitland, Maryland, outside Washington. The errata stated a mistake had been made in selecting the IBM punched cards. Instead of cards for the central United States, cards had been taken for halfway around the world in Soviet Central Asia but it was also a continental region (far from the sea) and the results were still valid. End digression.

While returning to the United States from across the Atlantic, I pondered why it took longer to fly from east to west, opposite to the direction of the earth's rotation, then from west to east. I had expected that the atmosphere would be dragged along by friction with the earth, with slippage, so I would have expected the opposite.

Why the Jet stream?

My explanation of the observation that the earth's predominant air flow is from west to east, with the atmosphere rotating more rapidly than the earth beneath it, is that there is an asymmetric solar pumping phenomenon with the morning sun evaporating clouds. The atmosphere applies a torque to the earth opposing the Jet streams. The flow of the atmosphere slows down by friction. Asymmetric pumping is a little like when you sit on a swing and, without touching the ground, move your feet back and forth appropriately during the swing cycle to swing higher and higher.

Columbus, while navigating toward the west, ran into trouble because his boat drifted out of the northern hemisphere trade winds. He did not have wind in his sails.

If you look up "Jet stream" you will see there are four Jet streams, two in the northern hemisphere, two in the southern. In each hemisphere there is a Jet stream in the temperate zone and one circling the Arctic/Antarctic, called the Polar Jet stream. It is very complicated. Jet stream winds can exceed 100 miles per hour (160 kilometers per hour). Flight routes are chosen to conserve fuel and reduce flight times. Jet streams do not go directly from west to east but also meander around mixing temperate zone air with Arctic/Antarctic air, warming the polar regions.

If the atmosphere were simply being dragged along by the earth as it rotates, there would be slippage and the predominant wind pattern would be from the east, not the west. The fact that the atmosphere rotates more rapidly than the earth itself means there is another mechanism at work besides the earth's rotation. Humidity evaporating in the morning expands the atmosphere. This causes the wind to come from the west. This mechanism provides the air mass momentum and an energy input from the west. This more than balances the surface air drag loss.

Another digression. Ocean currents affect the velocity of ships. Benjamin Franklin, sailing between North America and Europe, took a thermometer, attached it to a rope and dipped it into the sea. He was thus able to determine whether or not the ship was in the Gulf Stream being sped toward Europe or, on his return trip, if the ship was being slowed down. This method saved days on voyages.

Temperature influences

Temperature is a measure of the energy of moving molecules. In a room with sunlight coming through a window you can see dust particles bouncing around in what is called Brownian motion. The bouncing around is because the dust particles are being hit by more air molecules on one side than the other. In the air around us the molecules are moving faster than a jet plane. At typical room temperature the average air molecule is moving about 500 meters per second, approximately 1,000 miles per hour.

If you throw a ball up in the air, it slows down, losing the energy of motion, called kinetic energy. As wind is blowing upward against a mountain, it loses some of its energy of motion, and thus its temperature drops. The country, Ecuador, named after the equator, has a high mountain with a glacier on the top. Even there, the altitude difference causes a huge temperature difference.

When another meteorite like the one that exterminated the dinosaurs (except for birds) hits the earth again there will be a huge cloud of dust in the atmosphere, blocking sunlight from reaching the earth. The upper part of that cloud will be warm like the surface of the earth before that meteorite strike. The point is that the surface of the earth below it will be roasting, as in an oven, like the temperature difference between a glacier on a mountain top and sea level, not freezing. People have predicted that the surface of the earth would be expected to be freezing and not roasting.

Likewise, with soot in the stratosphere from many nuclear detonations on the earth's surface blocking sunlight from reaching the earth's surface, there hypothetically would be what was being called "nuclear winter" but would actually be "nuclear roasting." Thankfully, these tests were stopped by an international treaty.

The Hale-Bopp comet, visible from the earth in early 1997 without a telescope, was estimated to have a diameter of 40–60 kilometers (25–50 miles). Its next return is predicted to be in about the year 4385 and will miss colliding with the earth by a huge distance, if not changed by a collision in the Oort cloud beyond Pluto. If there is a future collision with the earth, the energy is estimated very roughly at 44 times that of the meteorite that killed off the dinosaurs. Smaller objects, like 150 feet across, will be far more frequent and when crashing into a deep ocean will create tsunamis with skyscraper-high waves rolling far inland in all directions. There is evidence that long ago sea bottom rocks were thrown up a steep cliff in the province of Victoria in southeast Australia. In some previous extinctions the meteorites were far more massive and the extinctions far more thorough than the one 66 million years ago. I was given a fossil of a trilobite (courtesy of my geologist daughter) that became extinct hundreds of millions of years ago after a massive meteorite strike which caused the climate change.

CHAPTER 8

∞

Slowing Earth Rotation – Why and Consequences

The earth is presently spinning at about 1,670 kilometers per hour (1,037 miles per hour) compared with about 600 miles per hour for a jet airliner. I quoted NASA figures, other sources state slightly different numbers.

The earth's rotation is slowing down; it is losing rotational energy because of tides, both with water sloshing around and tides within the ground itself. This decreased rotational energy is transformed into heat. It is like when you are driving down a steep mountainside and are applying the brakes and have shifted to low gear; the brakes are heated by friction and the engine, acting like a brake, gets hotter, getting rid of the increased heat through the radiator.

During the 1600s, Isaac Newton took a bucket of water and while turning his body he spun it around and observed the water climbing the walls of the bucket but not spilling out.

We know from earthquake shocks passing through the earth and bouncing off a dense core of heavy atoms like uranium, thorium, lead, and iron that the earth has a distinct core. The earth is heated by radioactivity in that core with a little more heat added by the slowing down of the earth's rotation. The earth's bulge is shrinking, but the shrinking lags the slowing.

Thus the earth's crust is cracked; continents and islands move. Examples include the Hawaiian Islands and the islands of Japan moving westward as the Pacific Ocean shrinks and the Atlantic Ocean widens. Lava flows upward through cracks in the earth's crust and volcanoes grow and sometimes blow off their tops causing earthquakes.

Periodically, a leap second is added to the coordinated Universal Time Clock (UTC) to compensate for the earth slowing down. The clock is at the Boulder, Colorado, facility of the National Bureau of Standards and Technology. Since the NIST-F2 cesium atomic clock was established in a chamber to be operated at minus 193 degrees centigrade, compared with absolute zero, the coldest it can get is minus 273.15 degrees, which is defined as zero degrees Kelvin. This atomic clock has been stated to be accurate to plus or minus one second in 300 million years. To me, the principal source of error will be the relativistic change in distance between the center of the earth and the altitude of the clock as the geology of the earth changes.

Leap seconds have been added periodically to keep the time correct. The first leap second was added in 1972. The most recent leap second was added on June 30, 2015, at 23.59: 60 before midnight making the year 2015 one second longer.

CHAPTER 9

∞

Some Quirks of History

Inscribed rock in Brazil

On a beach in Brazil, a rock was inscribed in old Hebrew, which was the regional language of both the Jews and the Phoenicians. It stated that, in the time of King Solomon, a boat left from what is now Eilat on a branch of the Red Sea, with a mixed crew of Jews and Phoenicians. It had sailed around the southern tip of Africa and up the coast, stopping to grow crops to feed themselves. It had been blown off course to the beach where the inscribed rock was found. What happened after that is unknown. The rock was placed in a Brazilian museum and stolen from that museum.

Pyramids

Pyramids in both Egypt and southern Mexico is no coincidence. There is evidence that ancient Egyptians traveled to Mexico and supervised the construction of the pyramids there.

Ancient influences

Rudyard Kipling (1865–1936) British short story writer and poet was born in Bombay/Mumbai in British ruled India. He won the Nobel Prize for literature. The reason I mention this is that he quoted British officers commanding the Indian Army as calling the Afghans that defeated them at the Khyber Pass "Jews" because their facial features strongly resembled the Jews they had seen back home in Great Britain. They did not know how right they were.

Seven centuries BCE, the Assyrian (not to be confused with Syrian) Army out of Nineveh conquered the Kingdom of Israel. Nineveh is now called Mosul. It is the second largest city in Iraq. It is in Nineveh province. The Assyrians had stopped killing everybody in lands they conquered and instead relocated them. The Second Book of Kings in the Bible states some of them were sent to Kabul, which is now the capital of Afghanistan, to guard the frontier of the Assyrian empire. Afghanistan is named after Afghan who was a descendent of King Saul of Israel, the King immediately before David.

The Pashtun tribe in Afghanistan and Pakistan has an ancient Jewish wedding custom. During the wedding ceremony, the bride and groom have a cloth held over their heads symbolizing the home. Without the recording of the oral tradition, two centuries after their own captivity during the Babylonian captivity, they were converted to Islam in the time of Mohammed, but many of them realize they have substantial Jewish ancestry.

After Babylon was destroyed by the Persians in a war in 539 BCE, Jewish merchants from Babylon established a trading post downstream from the ruins of Babylon. Over time it grew into the city known as Baghdad. Likewise, Fallujah was founded by Jews in exile.

Why is it USA, not USC after Columbus?

Why North and South America rather than North and South Columbia? Why are Native Americans called Indians after India? It is because Columbus knew, and was reminded, that the circumference of the world had been determined 18 centuries earlier by a Greek named Eratosthenes (276–194 BCE). Columbus chose to ignore this because financing and preparations for his voyage would have been endangered. He also had been in the Portuguese Cape Verde Islands off the coast of Africa and had seen driftwood on the shores so he knew there was land to the west.

Eratosthenes was a Greek living in Alexandria, Egypt. The ancient Greeks knew the world was round because the shadow of the earth on the moon during an eclipse of the sun—by the moon—was curved. He heard that one day a year the sun shone on the water at the bottom of a well to his south in Aswan, Egypt. He very cleverly erected vertical sticks where he was in Alexandria and waited for the day when the shadow was shortest. He measured the shadow, and from geometry, determined what fraction of the earth's circumference through the poles was the distance from Alexandria to Aswan. He inquired from caravans how long it took to travel between Aswan and Alexandria, and using the estimated distance a caravan traveled in a day, calculated the earth's circumference. In modern times, his estimate was determined to have been within 10%; the error being mostly in the distance between the well and him.

Columbus never formally admitted his deliberate fraud. Columbus chose the name Christopher (which in Greek means "messenger of Christ") to hide the fact of his Jewish ancestry. His father was a weaver in Genoa which was one of the few jobs Jews, who were not admitted to guilds for other employment, could hold. Some of his correspondence (surviving in Spanish archives) with his brother and others known to be Jewish show on their top little squiggles of the Hebrew letters *bet hey* meaning *baruch hashem* (blessed is the Lord). Also, in his will, he left one-tenth to a home for the aged for those forced to convert to the Catholic church (called Marranos, pigs in Spanish).

Amerigo Vespucci was an Italian from Florence. He worked in Spain on preparations for Columbus's first voyage so it is possible he met Columbus. He sailed for the kings of Portugal and Spain. His first voyage was in 1499, seven years after Columbus's first voyage. He discovered the mouth of the Amazon and sailed way down the coast of South America. After his return from his second voyage, he loudly proclaimed it was not Asia, and he had discovered a new land. A mapmaker put Vespucci's first name on the map, calling it America. When the coast of North America was explored and it was recognized it was a second continent, maps showed South and North America as separate continents.

America's founding fathers

George Washington, with the rank of colonel in the Virginia militia, served as an aide to British General Braddock during the French and Indian War. General Braddock was badly wounded and died. Upon his deathbed, General Braddock gave Washington his red sash signifying command. Later, in November 1758, George Washington led his troops into the smoking ruins of Fort Duquesne, renamed Pittsburgh, after British Prime Minister William Pitt who had pushed for support of the North American colonies against the French.

Washington and his troops at Fort Necessity were surrounded by the French, shooting from a forest, and forced to surrender. He soon escaped.

Later, during the Revolutionary War, Washington met with French military officials at Newport, Rhode Island, and in Connecticut. The French landed 6,000 soldiers at Newport to accompany Washington's army. They crossed the Hudson River upstream from New York using boats, around the clock, for three days. They marched south through New Jersey in a way that threatened the British occupying New York. The French landed an additional 3,000 troops near Yorktown on a peninsula in Virginia. A French fleet of 35 ships, outnumbering the British, landed siege guns and built fortifications surrounding the British.

The British were forced to surrender, which they did in style, with a marching band coming out of their fortifications playing the tune "The World Turned Upside Down." Then the surrendering British hosted the French and American officers at a dinner. Lord North, in New York, had ordered General Cornwallis to disrupt the revolution in the south. Lord North received the blame for the British defeat which ended the war. General Cornwallis received a promotion, being appointed Governor General of India.

The French fleet sailed from Yorktown to the Caribbean where it was destroyed by a British fleet, but too late. The British had already lost their 13 colonies, but retained Canada.

Who influenced the French to fight? Benjamin Franklin was sent by the Continental Congress to be their ambassador to France, leaving his wife behind in Philadelphia. His mission was to persuade the French to join the war against Great Britain. Benjamin Franklin was greatly respected in France for his invention of the lightning rod, used to protect their buildings against lightning, and a more efficient wood burning stove. He had earlier, flying a kite on a wire during a thunderstorm, proven that lightning was electricity. Others, trying to repeat his experiment, were killed by lightning. The French had heard of American frontiersmen wearing raccoon fur hats, so to humor them along, he ordered a supply of such hats from America even though he had never worn one.

The French were influenced to join the war efforts after the success of two battles when British General Burgoyne was forced to surrender south of Saratoga in October 1777. This was an American victory which helped Benjamin Franklin persuade the French to join the 13 colonies in the war against Great Britain. Spain also joined.

Among the American heroes at Saratoga was General Benedict Arnold who was wounded. Benedict Arnold was persuaded by his wife to switch sides and became a brigadier general in the British Army.

Why jeans are called jeans

Jeans were the name the French called sailors of the Genoa Navy and others from Genoa. During the gold rush in California a tailor named Levi Straus went from New York where his family had established a dry goods business to San Francisco. He took along a supply of brown sailcloth being used for covered wagons, etc. Miners were complaining their pants were quickly deteriorating, especially the pockets which held rock ore samples. He used his sailcloth for pants instead. From a sailing ship in the harbor of San Francisco which had come from Genoa, he purchased a supply of sailcloth, which was in blue, the color of the sailcloth and pants of the Genoa Navy; thus blue jeans. Years ago, before there were many competitors, blue jeans were commonly called Levi's.

Napoleon

Napoleon Bonaparte had a Corsican accent when he spoke French; an accent he never lost. It was related to the Italian accent of Genoa from which his family had moved.

Napoleon was born in Corsica, an island off the Mediterranean coast of France north of Sardinia. Genoa owned Corsica but was having trouble governing it, so it sold Corsica to France. World history was changed, including Napoleon's sale of what became the middle third of the United States (the Louisiana Purchase) extending from the Gulf of Mexico into Canada. Napoleon was anxious to sell it before the British seized it (as they had earlier seized Canada from France) because France could not defend it. President Jefferson quickly agreed, violating the Constitution by not getting the approval of Congress which might have been too late. Congress appropriated the purchase price of 15 million dollars in gold, paid twice, because the first ship sank on its way to France.

Jefferson appointed his secretary, Meriwether Lewis, to head an expedition Jefferson called "the Corps of Discovery" to explore the Louisiana Purchase. Lewis invited William Clark, who had been his commanding officer during the Revolutionary War, to be co-head. The expedition continued all the way to the Pacific, creating the basis for the United States' claim to what became the states of Oregon, Washington, and Idaho.

Also changing history was the formation of Belgium, splitting it off from France to weaken France after Napoleon's final defeat at Waterloo.

Robert E. Lee

I am very interested in the American Civil War for three reasons: First, I was born in 1925, 60 years after the Civil War ended and as a little boy my parents took me to Memorial Day and Armistice Day (now called Veterans Day) parades where many Civil War veterans were still among the marchers. Second, during the Depression, my father, a professional photographer, decided that if people could afford to go to Florida they could afford to have their pictures taken. In free outdoor concerts, our national anthem was played followed by "Dixie," the Confederate theme song during the Civil War. Most of the audience stood for both. Third, and most important to me, my high school principal, Mr. Longstreet. For most of my last two years of high school, he also taught a physics class. He told us he was a grandson of General Longstreet, Robert E. Lee's right-hand man during the Civil War. After I retired, I had time to study the Civil War and General Longstreet's role. During the Civil War, General Ulysses S. Grant commanded the Union Army that forced Robert E. Lee, commanding the army of northern Virginia, to surrender to him at Appomattox Court House (an historic site I once visited, taking my family) ending the Civil War.

Longstreet and Grant were close friends while at the army military academy at West Point and Longstreet was best man at Grant's wedding to Longstreet's cousin.

In a physics class, I answered a question from Mr. Longstreet as to which would protect a (mechanical) watch from a magnetic field—a brass case or an iron case. I said iron. Mr. Longstreet "corrected me," saying brass. I knew he was wrong. I did not dispute him; I knew it was much more important for him to keep his authority. He may have later checked the textbook.

It was 1943 and I was about to turn 18 in July and be subject to the World War II draft, so I did not apply to college. Mr. Longstreet checked which high school transcripts had been sent to which colleges for which students and noticed mine was not among them. He went out of his way to talk to my father about sending me to college and recommended where to apply. I followed his advice and was accepted at one of them where my application was not too late. A few weeks after I entered Union College I turned 18 and registered for the draft in Schenectady, NY. Because I was very nearsighted and the military did not want the risk of me shooting someone on our side, I was classified for limited service. This meant an office job but I could not type, or direct traffic, or get drunk soldiers or sailors back into their barracks and into bed. The local draft board where I registered had its own separate quota and a very large number of men in that category so I was never called up. Later, I felt I was about to be drafted. I responded to a navy posting for chemical warfare officers. I wanted to be on a ship rather than trudging through freezing mud. I walked to a navy recruiting station and signed enlistment papers. I was rejected, maybe because of my eyesight, or maybe because of my being a physics major and the navy wanted me to graduate.

During the Korean War, I was working as a civilian employee for the Potomac River Naval Command doing many varied tasks including: preparations for the Inchon landing, establishing a gun testing range, and testing our own designs and the guns we had obtained from a MIG-15 flown by a deserter from mainland China to Taiwan. I had learned to never be behind a gun or rocket under test because if it failed, parts would fly backward. The navy wanted to know the blast pressure downward on a ship's deck from the blast of the booster of a vertically launched Talos antiaircraft missile. So I went to the government contractor's site in the hills of West Virginia. A Talos missile booster was mounted horizontally ready for testing. One pressure gage was destroyed in a test. Another booster was mounted as a replacement. This time a porous oil filter was in front of the pressure gage trying to protect it while making the measurement. The back end of the booster blew off and it detached from the mount, traveling nozzle first, into a hill a mile away. I was standing about 12 feet to the side of the booster when it broke loose. Years later I was involved in somewhat risky flight testing of equipment. As a civilian, I was more at risk than many veterans who never left the USA.

Robert E. Lee was a son of Light-Horse Harry Lee of Revolutionary War fame. Robert E. Lee wife's mother was a descendent of Martha Custis, the widow who owned Mt. Vernon (a plantation on the Potomac) and married George Washington.

During the war with Mexico, which Robert E. Lee opposed as unjustified, he mapped a route through a lava field with sharp pieces of lava, to allow for quick capture of Mexico City. Thus "From the Halls of Montezuma" are the first words of the marines' hymn. Many officers on both sides in the Civil War met each other during the Mexican War.

The top general in the Mexican War (which added the southwestern third to the United States) was Winfield Scott, who though old and infirm was appointed by President Lincoln for a major role in the Civil War. Robert E. Lee had been superintendent of West Point for three years and earlier had worked for the army as an engineer on improvements to New York harbor, thus living in New York State.

Lee had opposed secession of states from the Union. In 1861, General Winfield Scott offered Robert E. Lee a major role. Lee turned it down, stating his first loyalty had to be to Virginia and he took command of the army of northern Virginia. In the 1863 Battle of Gettysburg, when Lee's army of northern Virginia had penetrated into Pennsylvania, Lee lost a quarter of his army. His right-hand man, General Longstreet, had advised him against the frontal assault on Union lines and suggested he get his forces between the Union forces at Gettysburg and Washington, threatening Washington so the Union Army at Gettysburg would suffer the greater casualties. Lee feared being trapped between two Union armies and declined to follow Longstreet's risky advice. We will never know what would have happened if he did. Robert E. Lee surrendered to Ulysses S. Grant in April 1865, then helping quickly end the Civil War. Robert E. Lee became president of Washington University in Virginia. After Lee's death, the University was renamed Washington and Lee. After Lincoln's assassination later the same month, his vice president became president, followed by Ulysses S. Grant.

America's national anthem and national march

Francis Scott Key wrote a poem that became the lyrics of the "Star-Spangled Banner" during the War of 1812 when he realized "that our flag was still there" at Fort McHenry in Baltimore Harbor. It was realized that the words had the tempo of a British drinking song written for a social club. Thus it is set to a tune of the enemy.

John Philip Sousa composed the words and music of "The Stars and Stripes Forever" which later became the official march of the United States. He was paid by Atlantic City, New Jersey, to conduct his band. This was an attraction to help make Atlantic City a great summer resort. When I was two years old and learning to walk on the Atlantic City boardwalk, I heard his band playing "The Stars and Stripes Forever." Hardly anyone knows the lyrics. A parody was written for a Marx Brothers movie, *Duck Soup*. During World War II, I heard sailors in uniform marching while singing, "Be kind to your web-footed friends because a duck may be somebody's mother."

The Roman calendar

The original Roman calendar was ten months, time enough for a woman to give birth. Later it was changed to 12 months. The 12th month was December (*dec* as in decimal).

Similarly, *sept, oct* and *nov* meant 7th, 8th and 9th but the Romans were used to the month names and the count being wrong didn't bother them. Julius Caesar proclaimed himself a god and renamed the fifth month, Quinilus, July after himself.

To keep the number of days in a year matched to the seasons, so July remains in the summer, he shortened February, which was popular with the Roman citizens who liked a shorter winter. His great nephew, Augustus, who ruled the Roman empire after a Roman civil war, followed his example—renaming the sixth month, to August after himself. The words tsar and kaiser are both corruptions of the name Caesar.

Tsars, Lenin, Stalin, Khrushchev

The tsars, to hold the Jewish population down, drafted men to serve in the army under Russian officers for 20 years. This backfired.

During World War I, the Germans sent Lenin into Russia to try to overthrow the tsar's regime. Much of the army, which hated the tsar, joined the Red Army which defeated the tsar.

Ioseb Jughashvili, one of the many names and nicknames Joseph Stalin went by, was born in Georgia in the Caucasus which was taken over by Russia in 1828. He was in training in Georgia to be a priest of the Georgian church, for which he received a scholarship. He became an atheist. There was speculation that he was recruited by Russia under the tsar to be a spy, infiltrating the Communists. He helped rob banks to help establish his credentials as a revolutionary. When Lenin took over the Russian government, he was on Lenin's staff. He changed his name to Stalin, which relates to steel in Russian.

Lenin left instructions that when he died he was to be buried. Instead, Stalin had Lenin's body preserved and put on display under glass in Moscow. He had his so-called friends, who knew him in his youth in Georgia, executed—all of Lenin's cabinet executed. Leon Trotsky, who under Lenin had successfully commanded the Red Army, leading it to victory, escaped to Mexico, but under Stalin's orders was tracked down and killed in 1940—stabbed with an ice pick.

Stalin was very suspicious and the Germans took advantage of this by planting rumors that his general staff, commanding the army, was plotting to overthrow him; Stalin executed his general staff. After the German attack on the Soviet Union, Generals Zhukov and Timoshenko were called back from the Far East where they had escaped Stalin's purge by being out of Moscow. They had commanded forces that had bloodied the Japanese in a border battle. Just when the Germans thought the war was about to be won because the Soviet Union (USSR) was using civilians, including women, to dig trenches on the outskirts of Moscow, trains brought 12 divisions of soldiers in white winter uniforms to Moscow on the Trans-Siberian Railway. They launched a surprise counterattack and the Germans were never again a serious threat to Moscow during the war. The Germans helped defeat themselves by issuing winter uniforms by their rulebook, on the same date as in Berlin which had a much milder climate; likewise, for antifreeze. In November, the radiators of German vehicles froze up and then leaked.

Stalin renamed Volgograd to Stalingrad after himself. During World War II, Hitler refused to allow a retreat during the battle for Stalingrad, partly because of the name, leading to the German Army there being trapped and surrendering. The original Soviet tanks from the early 1930s were very inferior to the German Tiger tanks. As the Germans were advancing in the Ukraine, the Soviets managed to load the machinery for tank production onto railroad cars and move it to Siberia where they put it in large log cabin buildings. They actually out-produced the Germans in tank and aircraft production and the Soviets succeeded to get a faster, well gunned design into production in Siberia using steel from Magnitogorsk in the Ural Mountains, well north of Stalingrad on the border between Europe and Asia.

The Soviets' production limitation was the lining for gun barrels. They lacked the minerals to produce wear-resistant steel alloy gun barrels. They heated the cannons to expand them, then quickly inserted the gun barrels. When the cannons cooled, the gun barrels were squeezed tightly and could never be removed. Cannons for battleships and other artillery were made the same way. The engineers of the USSR analyzed the problem and saw that a tank managed to fire two rounds on average before being put out of action. So they rifled (put spiral grooves) to cause the projectiles to spin for greater accuracy than ordinary steel gun barrels.

During the Korean War, when I worked at the Naval Gun Factory, my job included establishing a gun test range and testing both our own guns and Soviet guns of types they supplied the North Koreans—so I knew something about guns. The Soviet engineers were better than the German; they knew which surfaces needed careful machining and which did not. The Germans had trained apprentice machinists on noncritical surfaces just because that is the way they had trained them in the past. The Russians, from the days of the tsars, had dragged artillery through mud and seen to it that gun clearances were such that guns did not jam.

Nikita Khrushchev was among the successors to Stalin. He was from the Ukraine. Stalin collectivized the small farms for better control by his government, creating a famine during which Khrushchev's wife died. He hated Stalin and denounced him at a large closed meeting of the Communist Party but word of this leaked out. Friends of his had migrated to America before World War II when this was possible. He did not hate America. On a visit to America, he was deeply impressed that the cups for a drink were thrown out rather than washed and reused. In a speech at the UN, he took off his shoe and pounded it on the speaker's platform. His associates regarded him as undignified and removed him from office but I regard his action as similar to my own in college when I waved a safety razor around to be left alone to do my homework—much better than actually using weapons. Khrushchev retired to his villa on the Volga River countryside outside Moscow. His son, much later, came to the United States to teach at Brown University.

Franco

In July 1936, Spanish General Francisco Franco started the Spanish Civil War, completed in 1939, with a revolt in Spanish Morocco with his army there crossing over into the Spanish mainland. The Germans and Italians, practicing for World War II to come, sent military, including aircraft with pilots, to Spain, aiding Franco in taking over Spain. After the German *blitzkrieg* (lightning war) in 1940 and the German occupation of the French Atlantic coast, Hitler invited Franco to meet with him on the French Atlantic coast near the border with Spain. Deliberately, Franco's train was six hours late. The meeting lasted nine hours. Hitler demanded the right to air bases in Spain from which German aircraft could attack ships, closing British resupply to its military in Egypt and through the Suez Canal beyond. Franco refused, as too risky, and said maybe later.

Franco agreed to Hitler's demand for lists of Jews in Spain to be rounded up for extermination at a later date. Franco also demanded Gibraltar and part of then French North Africa after the war was won. To help appease Hitler, Franco agreed to send a division of volunteers, called the Blue Division, to the Russian front. Hitler did not want to upset his timetable for a planned invasion of the USSR.

After his meeting with Franco, Hitler told Mussolini, "I would rather have teeth pulled by a dentist than meet with Franco again."

The British warned Stalin that their intelligence reported German preparations for an attack but Stalin refused to believe them and was caught by surprise.

After the German invasion of the Soviet Union bogged down and the outcome became uncertain, Franco wanted Spain to remain neutral in World War II. In the Soviet Spring Offensive following the German defeat at Stalingrad, the forces of Germany's allies, including the Spanish Blue Division, were wiped out and the Germans fled the Caucasus as fast as they could to avoid being cut off and captured. The roundup of Jews in Spain was never implemented.

To not appease Hitler was very dangerous. King Boris of Bulgaria met with Hitler in Germany but did not agree to Hitler's demands which included the handing over of Bulgarian Jews for extermination. On the flight back to Bulgaria he was poisoned and died two days later. The 48,000 Bulgarian Jews who had citizenship survived World War II. 11,000 non-citizens were handed over to the Nazis.

Later, a Jewish organization opened a Swiss bank account for Franco in case he was ever ousted from Spain and would need the money. In return, Spain allowed any Jews who reached the Spanish border to enter and pass through Spain. In 1942, my future wife and her mother, sisters, and brother-in-law left the German occupied Netherlands and smuggled themselves to the Spanish border. They entered and passed through Spain on their way to Portugal where they boarded a Portuguese ship to the then Dutch colony near the equator on the Atlantic coast of South America, now an independent country called Suriname. Thus they survived.

What Hitler did not know was that the Spanish name Franco meant someone who had come from France, mostly Jews fleeing the church in France before the Spanish Inquisition.

Two Tripoli's, two Galicia's, two Brooklyn's, two Harlem's

The Tripoli which is the capital of Libya, and mentioned in the hymn of the United States Marines: "to the shores of Tripoli" was founded in antiquity by Phoenicians from Tripoli in what is now northern Lebanon.

Galicia, Poland, was named after Galicia, Spain, to help make a Spanish princess feel welcome when she married Polish royalty.

Brooklyn (*Breukelen*) the borough of New York, and Harlem the neighborhood of Manhattan, were both named by the Dutch settlers of Neiuw Amsterdam (later taken over and renamed New York by the British) after towns in the Netherlands.

Commodore Perry's visit to Japan – side effects

In 1852–53, Commodore Perry forced a reversal of Japanese policy which had been to isolate itself from the rest of the world. He commanded a squadron of steam powered warships. Coming into a Japanese harbor, he demanded that the Japanese accept a letter from President Fillmore about opening trade between the United States and Japan. When the Japanese refused, he directed cannon fire that destroyed buildings on the shore. The Japanese accepted the letter. His squadron then sailed to China, later returning to Japan. The Japanese realized their policy of isolation must be reversed or they would become a colony of a western country as India had become a colony of Great Britain and the Indies had become the Dutch East Indies (now Indonesia).

They accepted President Fillmore's demands and started to industrialize. They sent young men to America to work in factories and learn industrial skills. In 1906, they defeated Russia in a war over Port Arthur in China. A later side effect of their industrialization was that it helped lead to the attack on Pearl Harbor and America's entry into World War II.

Consequences of an accidental explosion

The American battleship, Maine, making a good will visit to Havana harbor in Cuba, then a colony of Spain, blew up and sank. Spain was blamed for attacking it, especially in the newspapers of William Randolph Hearst. The United States declared war on Spain in 1898. After World War II, the sunken wreck was examined and it was determined the ship was sunk by an accidental explosion inside. Consequences included the charge in Cuba on San Juan Hill for which the well-publicized Theodore Roosevelt, leading the Rough Riders, received much of the credit. The United States retained Guantanamo Bay as a naval base and gave the rest of Cuba its independence. Theodore Roosevelt who had become governor of New York became McKinley's running mate and then vice president when President McKinley was reelected. McKinley was assassinated and, at age 42, Roosevelt became president. The Roosevelt family name became famous helping his fifth cousin Franklin Delano Roosevelt (FDR) establish himself in politics, becoming governor of New York and later president of the United States. He married another fifth cousin, Eleanor, with President Theodore Roosevelt serving as best man. The two Roosevelt families knew each other well.

The population of the United States was very isolationist after the casualties of World War I. FDR was very concerned that the British might be starved out and forced to surrender by the German submarine fleet sinking of British ships, some of which were torpedoed in shallow waters off the south Florida coast. The Civil Air Patrol of volunteer pilots with their planes was spotting submarines in shallow waters against a sandy sea bottom background. I heard depth charges detonating. Oil tar from sunken tankers rounding Florida from the Gulf of Mexico was washing onto south Florida beaches. I stepped on some and got a rag and some kerosene from a lifeguard to rub tar off my feet. This was after World War II had started in Europe and about two years before the Pearl Harbor attack brought the United States into the war.

Roosevelt got Congress to approve Lend-Lease under which the United States lent Great Britain 50 destroyers to be manned by British crews to guard convoys. In return, under what was erroneously called "Lend-Lease" the United States obtained bases in Bermuda, the Bahamas, Jamaica, and Trinidad.

Code breakers broke the code of the German ultra coding machine. An American team of code breakers broke the Japanese code; this was kept a high priority secret.

FDR was very concerned that the Axis (Germany, Italy, Japan) would take over the rest of the world including Latin America and we would be alone fighting the rest of the world. He arranged with the British who controlled Middle East oil and the Dutch government in exile who controlled the oil fields of the Dutch East Indies to serve an ultimatum on Japan in June 1941 to withdraw from China or face an embargo of shipments to Japan of oil, scrap iron, etc.

At that time, Japan was prepared for such an embargo and had an 18-month supply of oil in storage. Roosevelt and Churchill knew the Japanese military could not accept this ultimatum and would attack, with the United States thus in World War II. There was concern that Germany would not be at war with the United States but Hitler quickly cooperated by declaring war on the United States in solidarity with Japan. The war in Europe (and North

Africa) took priority over the war in the Pacific. Admiral Nomura, who commanded the Japanese forces attacking Pearl Harbor and had been educated in the United States at MIT, warned the Japanese government, "Do not kick a sleeping giant." Both the Japanese and the Americans were incompetent on Sunday, Dec. 7, 1941.

The Japanese failed to attack the oil storage tanks at Pearl Harbor. They also did not attack and destroy the floating dry-dock that was used to repair damaged ships. The Japanese fleet maintained radio silence, signaling ship to ship with lights, and approached north of the Hawaiian Islands avoiding the southern route where they could be seen by commercial traffic between Asia and North America. Their biggest problem was that America's aircraft carriers were safely out to sea on training exercises on Dec. 7, 1941. The Americans detected and sank a mini-submarine, launched from a larger submarine, trying to enter Pearl Harbor. Before the dawn attack on Pearl Harbor, the second mini-submarine penetrated into Pearl Harbor and, firing its torpedoes, achieved much of the damage attributed to torpedoes dropped from low-flying planes. Those who sank the mini-sub did not phone the commanders in Hawaii and wake them up.

Also, a radar station was established on the north coast of Oahu to detect any incoming aircraft that might be attacking and provide precious minutes of warning time of a possible attack. A large group of approaching aircraft was detected by the radar operator who informed his commanding officer at the radar site. The officer assumed it was American planes returning from a training flight and failed to phone an alert. That assumption voided the whole purpose of the radar site. He was court-martialed, imprisoned, and later expelled from the military with a dishonorable discharge. Army Lieutenant General Walter Short and Admiral Husband Kimmel, being in overall command, were blamed for the failed defense against the

Japanese attack. The warning they received had few details and was misinterpreted as a possible attack by some of the large local Japanese population on Oahu, so they moved the aircraft away from the edges of the main airfield on the island and grouped them closely together making them better targets for the attack from the air.

The Japanese code had been broken. Information about the attack on Pearl Harbor was understood. There was confusion because the Japanese embassy in Washington had not officially declared war as had been requested in the decoded message. The time zone difference between Hawaii and Washington caused this delay and therefore the threat was misinterpreted.

On June 4–7, 1942, the Japanese fleet including four aircraft carriers attacked the American base on Midway Atol. Having broken the Japanese code, the Americans knew what the Japanese were saying to each other in ship-to-ship radio broadcasts. The Americans, with three aircraft carriers, lost one while the Japanese lost all four, ending the Japanese numeric superiority in the Pacific.

The day after the attack on Pearl Harbor, the Japanese, flying planes from other aircraft carriers, destroyed the unprepared American planes in the Philippines under the command of General MacArthur.

On a personal note, many years later I was on a business trip to Washington and extended it into the weekend, staying with my wife's family. That weekend I was at Arlington National Cemetery with others for a memorial service for one of the team that broke the Japanese code as her ashes, in a tray, were being placed in one of many slots in a stone monument. I had met her over the years at various times but did not know her as well as my wife's family did.

Arlington National Cemetery was on the property confiscated from Robert E. Lee for his actions as commander of the army of northern Virginia during the Civil War.

President Roosevelt was very careful not to leave a paper trail of how he had succeeded in getting the United States into World War II. He was concerned that if America did not enter the war, it would have to face a world dominated by the axis, alone. I rank him as one of America's greatest presidents along with Jefferson, Lincoln, and Truman. George Washington was great, but his greatest accomplishments were leading to victory in the Revolutionary War before he became president.

Assassinations, failed and completed

Abraham Lincoln, because of rumors about an attempt to assassinate him as he passed through Baltimore on his way to Washington to be inaugurated as president, followed security advice to wear a disguise and have his special train pass through Baltimore at night.

In 1933, president-elect Franklin Delano Roosevelt spoke in a Miami public park. A short, deluded man, who had emigrated from Italy and became a U.S. citizen, stood on a folding chair to look above the heads of the crowd and have a clear line of fire. He held a pistol he bought from a pawn shop and fired it at FDR. The chair wobbled and, instead, the bullet killed the mayor of Chicago, sitting next to FDR in an open convertible. The assassin fired wildly at the crowd and was himself shot. He was tried, convicted, and electrocuted.

In March 1933, Roosevelt was inaugurated. The inauguration was then still in March because in the early years of the United States travel in the winter to a central location was too difficult—later changed to January to reduce the time between the November election and those elected taking office.

In September 1901, early in his second term, President William McKinley was shot by an anarchist and died the same month. His anarchist assassin was electrocuted. He was at the Pan-American exhibit in Buffalo, New York, when he was shot as he reached out to shake hands with the man who was about to shoot him. His secretary had been concerned for his safety and had twice removed his greeting the public from his schedule, but McKinley overruled him each time. His vice president in his first term had died.

Theodore Roosevelt, a hero of the Spanish-American War in 1898, had been elected governor of New York State. A political boss wanted to get rid of Roosevelt in the state and helped arrange for him to be nominated as vice presidential candidate at the Republican nominating convention. Theodore Roosevelt was a popular president. He established the National Park Service, soon copied by Canada and then many other countries. Because of the Roosevelt name, President Wilson, a Democrat, appointed him undersecretary of the navy to run the navy under a political figurehead during World War I. (Wilson himself had been president of Princeton University where some alumni wanted to get rid of him and arranged for him to run for governor of New Jersey, from which he went on to become president of the United States.)

President John Kennedy said "life is not fair" a few weeks before his own assassination. The commission headed by Chief Supreme Court Justice Earl Warren investigated and issued a long report blaming Lee Harvey Oswald alone, firing from above and behind the convertible car Kennedy was riding in. This conclusion was disputed, I believe correctly. A man in the crowd, Abraham Zapruder, was filming a movie of the scene and it shows Kennedy's head being shoved back, not forward, as the bullet hit, showing convincingly, that the bullet came from in front of the car. You can see this on You Tube. Who was responsible is a question and there are many conspiracy theories. No other person was caught and convicted.

The 1930s and '40s

I was born in 1925 while Coolidge was still president. I lived through the Great Depression and on my way to school read the headlines on outdoor newsstands about Hitler coming to power and Franklin Delano Roosevelt being elected. Radio broadcasting began a few years before I was born and I took it for granted just as today's young adults take TV broadcasting for granted although it did not begin until after World War II when I was already an adult. On the radio I heard Adolf Hitler speaking with a running translation. I did not need to hear the translation. From the tone of his voice I knew great trouble was coming.

Why the Netherlands was invaded during World War II

After Germany's defeat in World War I, the Kaiser was forced to go into exile. He went to the adjoining Netherlands. In his retirement, he chopped firewood for exercise. Germany was prohibited from building aircraft and the German aircraft company, Fokkers, opened a factory in the Netherlands where it built aircraft and bicycles. Originally, Germany had no intention of invading the Netherlands in World War II.

The Munich Agreement under which Czechoslovakia ceded the largely German populated Sudetenland to Germany was signed September 30, 1938, making Czechoslovakia defenseless. British Prime Minister Neville Chamberlain said, "Peace in our time," but actually knew better. He ordered speed-up of development and deployment of Great Britain's radar, command and control, and fighter plane defenses against the expected German bombing. An important procurement was variable pitch propellers which function like a gear shift in an automobile. They allowed the British fighter planes to climb and descend faster, and by conserving fuel, to stay in the air longer before landing to refuel; these were purchased from Hamilton Standard in Connecticut.

One quarter of the German armor for the *blitzkrieg* to the English Channel was taken from the Czechoslovak Army and the Skoda Works in Prague became a major arms manufacturing facility for the German war effort.

On September 1, 1939, Germany invaded Poland starting World War II. After the *blitzkrieg*, Germany quickly occupied Denmark and flew aircraft with soldiers into Oslo Airport. The objective was to establish submarine bases in the fiords of northern Norway so the British could not keep the German submarine fleet bottled up and Germany could force British surrender by starving them out. The Dutch learned from Norwegian experience and erected barriers at Amsterdam's airport that wrecked German transport planes as they tried to land soldiers. These same planes had been used to train pilots, creating a pilot shortage. By contrast, the United States employed Link flight simulators to reduce training losses but still lost more pilots in training than in combat.

I read long ago that in November 1940 a high ranking German general was being flown as a passenger over the Rhineland when a heavy fog set in. The pilot flew low so he could see the ground and found a field he could land on. Upon landing, a farmer came out to greet them. They found out they had landed across the border in Belgium. The general did not smoke. He asked his pilot for matches or a cigarette lighter but the pilot did not smoke either. Likewise, the farmer did not have matches on him. The farmer escorted them into his farmhouse. It was a chilly day and the farmer had a stove with embers in it. The general reached into his briefcase and took out papers that he thrust into the stove. At that moment a Belgian policeman arrived, saw what was happening and grabbed the papers before the embers could ignite them. Those were the plans for the

German *blitzkrieg*, except for one detail that had an even higher security classification. Thus the Germans revised their plans to include going through the Netherlands. The higher classified detail was dropping parachute troops who would steer themselves to land on top of pillbox fortifications and throw hand grenades through openings into the pillbox fortifications below them.

Nine years later, I met and married my darling wife in Washington DC. She was then working at the Netherlands embassy; I was working as a physicist at the Naval Research Laboratory. Her mother, by then a widow, had led her family out of the German occupied Netherlands in July 1942 to Suriname. During the war it was protected by American as well as Dutch military because it was a source of bauxite (aluminum ore). She had completed her high school-plus education and replied to a posting by the Netherlands embassy in Washington—it helped that a sister had preceded her. She took a bauxite boat to Mobile, Alabama, delivering ore to be processed into aluminum by the Tennessee Valley Authority. She was met by the honorary Dutch consul who, the next day, put her on a train to Washington. She was still a teenager, turning 20 later that month. She had gotten her job because she could work in Dutch, English, French, and German. Many years later at a ceremony for a granddaughter of ours, she also spoke, carefully rehearsed, in Hebrew, but could not carry on a conversation in Hebrew.

Years after we were married, my wife told anecdotes of what happened during the German invasion of the Netherlands and afterward. The Germans were bombing the heart of Rotterdam to intimidate the British before the coming Battle of Britain. This backfired, making British preparations more urgent. After the war, in the center of Rotterdam, a metal statue was erected of a man with an opening where his chest would normally be. The Hague, where my wife's family lived, was close enough to Rotterdam to hear the bombs exploding. In case The Hague would be included in the bombing, my wife's family went into a basement for some protection in case the roof collapsed. While in the shelter, my wife's aunt used the time to put curlers in her hair.

Later, the Jewish students were forced out of public schools and put in a segregated school with Jewish teachers. The young girl I later married was, like others, ordered to turn in her bicycle for German soldiers to use. She had been complaining to her parents that she had outgrown it. When she went to turn it in, it was refused because it was too small. When she road it to school, her sisters organized a cheering section. There were about 30 students in her class and after the war it was determined that four somehow survived.

I should add that not all the Nazi soldiers in the Netherlands were bad. One Austrian Army officer was very helpful to my late mother-in-law, by then a widow. He purchased the stock of the family shoe store downstairs from where they lived, leaving empty boxes on the shelves, and shipped the shoes to Vienna to be sold on the black market there. This provided funds to help finance the family's escape from Hitler's Europe. He was careful to never give his name. Also, in July 1942 on the day when Jews in Amsterdam were rounded up for cremation in facilities built in Poland, he warned her that the round up where they lived would be the next day so they left quickly, with many preparations for the dangerous trip already made. Several times they played "Russian roulette" and won.

My late mother-in-law was born in the Russian occupied part of Poland and grew up in Germany. She lived in Germany during World War I and spoke an excellent German which helped in her relations with the Austrian officer. After World War I, but long before Hitler came to power, the shoe store she and her husband owned in Bremen had its windows broken and was looted by anti-Semites, so they moved across the border to the Netherlands where my wife was born. Hitler's rabble-rousing fell on fertile ground.

Adolf Hitler

The name Hitler, adopted by some ancestor, is a variation of the Czech *Heidler*. As a young boy his father severely beat him, probably suspecting he was not his biological son. Neighbors said Adolf Hitler was different than his sisters because he had some talent in contrast to his sisters who were considered simple-minded. Adolf Hitler was partially the product of an Austrian Catholic education and environment which was viciously anti-Jewish; consider the passion plays reenacting the crucifixion. As a young boy he played pope and got other young boys to kiss a ring on his finger. He also feared he might be "contaminated" by some Jewish blood and his actions were possibly, in part, a denial of this. In World War I, he served in the army under a Jewish sergeant who recommended his promotion to corporal.

Hitler complained that Germans were a minority in Austro-Hungary. Franz Joseph, the emperor, was very popular with all the different ethnic groups in his empire and he spoke many languages: German, Hungarian, Czech, Polish, Serbo-Croatian, and Yiddish.

After World War I, Hitler was largely unemployed and living in a homeless shelter funded by wealthy Jews who were giving back to the Vienna community. Hitler was very resentful of having to depend on Jewish charity. This illustrates with a vengeance the old saying "no good deed goes unpunished." Adolf Hitler cried on the shoulder of a Doctor Bloch who was with Hitler's mother as she died of cancer.

After Germany took over Austria in 1938, with a big welcoming ceremony, Hitler intervened to let Doctor Bloch and his family leave Austria. Another exception he made to Germany's anti-Jewish laws was for Nobel Prize winner, Otto Warburg. Hitler was extremely afraid of cancer. He removed Warburg as director of the Kaiser Wilhelm Institute but had him work on cancer research. Hitler also hired a Jewish woman to be his acting coach to help him overcome his very shy nature and be forceful.

The actor, Charlie Chaplin, called Hitler a great actor. After the *blitzkrieg*, Chaplin made a movie, *The Great Dictator*, ridiculing Hitler and Mussolini. A Hungarian Jewish second hand clothes merchant unfortunately took pity on Hitler when he lived at the Vienna homeless shelter and gave him an unsaleable but warm coat so he would not catch pneumonia in the winter. If he had died, then history would have been very different. If he had not committed suicide before the occupation of Berlin by the advancing Soviet Army and been put on trial with an assigned defense attorney, the defense could have been "not guilty because of insanity," a judgment that could also be applied to many others.

Hitler's body was taken outside after he shot first Eva Braun and then himself. His body was then doused with gasoline and set on fire. It was identified by comparison with his dental records. I offer a very speculative plot that I do not believe but it makes a good story. He was not that clever and it would have required cooperation by his dentist. The dental records of another patient very similar to his own would have been labeled as Hitler's and that person's body, shot through the head as in a suicide, substituted for his own. Meanwhile Hitler would have been in hiding in an apartment in Berlin while he grows a beard and is circumcised. Then he leaves the apartment for a displaced persons camp where he meets Jews and learns from them a few words of Hebrew and English and names of some Jewish holidays and what they are. He is assigned to a kibbutz where he receives work assignments, learns Hebrew and lives out the rest of his life there. Not impossible but very unlikely. Fits in with his having taken advantage of Jews before, from the corporal who recommended him for promotion during World War I to the Jewish acting coach who taught him to practice being forceful, in front of a mirror, to overcome his naturally shy personality. Hiding out as a Jew would not be expected of Hitler and he probably would have gotten away with it.

Neither Hitler nor Goebbels, his propaganda chief, looked like the tall Aryan warriors, like the Vikings Hitler praised. They were short, and Goebbels invented a second type of German: short elven-like, dwelling in the forest.

Hitler patted a blond, blue eyed boy on the head, not realizing the boy was Jewish; somehow the boy survived the Nazi regime and made it to Israel where he told the story years later.

How Switzerland avoided being invaded

I remember that before the German *blitzkrieg* across the English Channel there was speculation as to whether the Germans would outflank the French Maginot Line and its extension, the Belgian Leopold Line, by going through the Netherlands or Switzerland. I learned, partially from a good friend and coworker, who with his wife knew how this was accomplished. The Swiss had drilled holes in the ceilings of the railroad tunnels under Swiss mountains and placed explosives in the holes, ready to detonate. The railroads connected Germany with Italy. Also soldiers like my friend took their rifles and ammunition home with them; there were no armories that could be quickly seized to disarm Switzerland. Then the Swiss invited the general staff of the German Army to inspect the Swiss preparations. The German general staff reported to Hitler that they estimated it would take ten years to reopen the tunnels and Germany would lose a half million men in an invasion of Switzerland.

So Hitler decided that he would wait until the war had been won and Switzerland, being surrounded, would have to surrender. Also, many in Hitler's regime had opened Swiss bank accounts in case Germany lost the war. Many German Jews had also opened Swiss bank accounts and when they did not survive, the Swiss banks conveniently "lost" the records and kept the money.

The arrangement between Switzerland, Germany, and Italy was that part of the Italian port of Genoa would be turned over to Switzerland for exports and imports and Switzerland would continue trade with Germany, including selling Oerlikon (named after a suburb of Zurich) antiaircraft/antitank/anti-surfaced submarine guns to Germany. Oerlikon guns were built and used by both sides, including even on Japanese Mitsubishi Zero fighter planes in World War II.

Adolf Hitler chose not to be present at the January 20, 1942, Wannsee Conference in the Wannsee Park in Berlin but he undoubtedly gave spoken commands to those who ran the conference as to what the decisions would be. Champagne and other refreshments were served. After World War II, the British, who were the occupiers of the Berlin section including Wannsee Park, discovered the carefully recorded minutes of the conference meeting. I read the English translation of this record, published in the American edition of the British science magazine *Nature*. These minutes reported the decisions to proceed with "the final solution" to exterminate Jews, gypsies, and homosexuals.

To make space for Germans to live in Poland and the Ukraine, the ethnic Polish would be gradually exterminated but this would have to wait until the war was won. In practice, the Germans proceeded to murder Polish intellectuals such as college professors, even those who agreed with them about killing Jews. The German Army was already in trouble at Stalingrad and it was otherwise too risky to proceed with cremating the mass of the Polish population.

People were transported in boxcars from western Europe across Germany for extermination at Auschwitz and other cremation centers. It was understood that the U.S. State Department and the American military under Chief of Staff General George Marshall did not want to discourage what the Germans were doing at Auschwitz and gave unofficial orders not to bomb the railroad yards at Auschwitz.

When railroads were about to be built in Russia, the Russian government (under the tsars) wisely directed that the spacing between the rails be wider than in western Europe so the railroads could not be easily used in an invasion. During World War I, at Auschwitz, trains from Germany were unloaded and then reloaded onto trains built with the wider rail spacing.

My cousin, Jack, was a bombardier-navigator using a Norden bombsight on a B-17 based in northern Italy. His crew received orders to bomb the railroad yards at Oswiecim in Poland. They did. On a later flight, their plane was hit by antiaircraft fire and caught fire. The pilot, captain of the flight, ordered the crew to bail out.

As Jack landed in Germany and was getting out of his parachute harness he was captured by a farmer with a pitchfork. The tail gunner had been wounded and the pilot put the plane on autopilot and went back to help the tail gunner. The plane climbed to high altitude and in rarified air the flames went out; the pilot flew the plane back to base in Italy. It was a year before Jack's worried parents, my Aunt Rose and Uncle Sam, heard through the International Red Cross that he was still alive and a prisoner of war (POW).

The prison camp was in Poland. There he met a friend from his high school class and after the war they had lunch together or otherwise met about once a week. His friend was at his funeral. As an officer, Jack was given preferential treatment, but was still starved down to less than half his weight. His brain shrank with the rest of his body, resulting in headaches the rest of his life. His metabolism permanently adjusted to starvation and he gained weight and in his last years was on dialysis.

Like others in the military, he wore a so-called "dog tag" which identified his body in case of death or serious injury, for purposes such as Catholic last rites. His "dog tag" identified him as Jewish. The two men who ran the POW camp had been communists when Hitler came to power. A few months before the war ended in Europe, Hitler gave orders to execute Jewish POWs. They chose to ignore Hitler's orders. The Soviet Army advancing toward Berlin liberated the POW camp. Then Jack found out Oswiecim was the Polish name for Auschwitz. A few years later Jack was best man at our wedding and later I was at his.

Fritz Haber

Fritz Haber was a Jewish chemist in Germany. He invented a process to take nitrogen out of the air, which is about four-fifths nitrogen, and reacting it with hydrogen under temperature and pressure in the presence of a catalyst to produce ammonia. The ammonia, in turn, is used to produce fertilizer and explosives. This allowed Germany to fight World Wars I and II without imports of nitrates for explosives and in spite of blockade by the British. He won the Nobel Prize for chemistry in 1918. When Hitler came to power, Haber was in deep trouble. He died in Switzerland in 1934 while in transit to Rehovot in Palestine (now Israel) for a job as director of what would become the Weizmann Institute, as arranged for him by Albert Einstein.

Eisenhower

As the Allied landings in Europe were being planned, an American general was appointed to be in overall command.

The plane he was flying crashed in bad weather while landing in Iceland for refueling on the way to England. Today who knows his name? Dwight Eisenhower, a career army officer, was promoted and sent as his replacement. Hitler stated that the Americans had to appoint an ethnic German to be their commander; Eisenhower translates into English as "blacksmith." His family were farmers in Kansas.

I will not repeat the well documented history of World War II in Europe except for a few remarks. The Germans complained that it did no good to know the American plans because the Americans did not stick to them. After the Americans landed in North Africa, German spies learned that Roosevelt and Churchill were going to meet at Casablanca in Morocco. In Germany, this was not believed both because it was so soon after the landing and because Casablanca translates as "white house," so they thought it would be at the White House in Washington.

The Germans had spies counting the planes coming out of American aircraft factories. The counts were accurate but the Germans refused to believe them, regarding them as impossible.

After the war, Eisenhower received the Republican nomination for president and became the next president after Truman. His vice president was Nixon. Eisenhower's son married Nixon's daughter. Eisenhower had a stroke while in office and had to relearn how to speak.

At Columbia University the alumni council was seeking the next president of the university. One of the council members knew President Eisenhower's younger brother, Milton, was a great educator at John Hopkins University so he nominated him but did not state his full name. The other trustees jumped at that and chose retired General Eisenhower.

My wife and I knew a female professor at Columbia. We asked her what she thought of Eisenhower as Columbia's president. She did not know his accomplishments as an administrator or fund-raiser but said he had ordered sidewalks to be installed where students were taking shortcuts across the grass. After Columbia, Eisenhower bought a farm near Gettysburg and lived the rest of his life there.

The Norden bombsight

The Norden bombsight got great publicity during World War II. Its accuracy was exaggerated with statements like placing a bomb in a pickle barrel. Uncertain wind between where the bomb was released and where it landed limited accuracy. My cousin, Jack, as a B-17 bombardier-navigator used a Norden bombsight. During the bombing runs he took control of the bomber, steering the bomber left-right until the bombs were dropped. I worked on the design of successors to the Norden bombsight, incorporating radar so they could function at night.

I never met Carl Norden but knew engineers who had worked for him. Carl Norden was born in Holland. My wife, who was from Holland, told me that north in Dutch is *noord*—Norden is a variation. He was trained as an engineer in Switzerland. He received a contract from the navy in the 1920s for a bombsight and designed it in Switzerland while on summer vacation. He sent the drawings back to his assistant, Ten Bosch, who had it built at Carl Norden's company in lower Manhattan. I heard that Carl Norden had a German employee at his company who gave a copy of the design to Germany long before World War II. Neither Germany nor Japan, having their own bombsight designs, ever chose to build it. As a designer of its successors I state that the principal error was introduced by the unknown changes in the wind velocity along the bomb's path toward the target on the way down.

To have enough production, the Norden bombsight was also built by Bendix and by a navy owned facility in Indianapolis. After the war, the federal government, under a law tried to claw back excess profits from companies. Escaping this, Carl Norden went to Switzerland and lived the rest of his life there.

In the base of the Statue of Liberty is a small museum showing the accomplishments of immigrants to the United States. We visited it. A Norden bombsight was prominently displayed.

Arrangements for occupation of Germany

Roosevelt, Churchill, and Stalin met at Yalta in the Crimean Peninsula of the USSR as the Americans and British were advancing from the west and the USSR was advancing from the east toward the occupation of Germany. They defined the areas to be occupied by each, with Berlin well within the Soviet zone (which was to be divided into three zones).

When the American and Soviet armies met in Germany, the Americans did not speak Russian and the Russians did not speak English but they needed to coordinate. Their common language was Yiddish.

To hold down casualties of the American and British soldiers under his command, Eisenhower slowed their advance. He stated "five Ivans are dying for every John." Meanwhile Stalin sent his army into Berlin taking very heavy casualties before raising the Soviet flag over the Reichstag, the German capitol building. Casualties could have been greatly reduced by starving the Germans into surrender.

Vietnam – the unnecessary war

The United States was concerned with the spread of Communism. This was a reason for sending soldiers to

Vietnam. If the United States would have noted that the North Vietnamese resented the Chinese Communists for having occupied part of their country, then maybe this war could have been averted. The North Vietnamese were not pro-Chinese as the United States had thought. Communism from China would not have spread into North Vietnam.

Why are you who you are?

As an example, I use myself. I heard from my parents and other relatives many stories of family history. About 1902 in New York my father's father was laid off as a typesetter from a newspaper when it was about to start using linotype machines, with younger men to be trained to use the linotype machines. He answered two ads he had just set for little general stores for sale like the one he had previously run with his wife in Russia. One was in Yonkers, New York. The other was in Newark, New Jersey. He bought the one in Newark where my father, much later, met and married my mother. Otherwise my genes would be split between two individuals who would probably never have known of the other's existence.

How did your grandparents meet? Where did they come from? Likewise, your parents. Think about this.

A humorous story to conclude this chapter

We were transferred to Connecticut in 1953 by my employer to help meet commitments on projects for our national defense. At that time, trains were boarded and exited through a flight of stairs from the passenger cars to the ground just above the level of the railroad tracks. At Grand Central Station in Manhattan, the conductors on the passenger cars would fold down a metal shelf to be level with the passenger car floor and the raised platform the passengers would be stepping out onto. The New York, New Haven and

Hartford Railroad ordered a modern looking low-slung train, I believe from Bombardier in Montreal. Passengers on its maiden trip from Hartford to Grand Central Station in Manhattan included the governor of Connecticut and other politicians, as well as railroad executives. When the train arrived at Grand Central Station, the train doors could not be opened and the heads of standing passengers looking out were a little above the floor of the station platform. The railroad sold the very modern looking train at a loss to the Spanish railroad system and there it was called the tango; a humorous fiasco.

Some loose ends

A cartoonist, James Montgomery Flagg (1877–1960) drew a figure of a tall, slim man with red, white, and blue clothes and called him Uncle Sam (for United States). On his poster were the words "Uncle Sam wants you" and a finger pointing at the reader. It was used as a recruiting poster for World War I. During World War II, as German submarines were sinking ships off the coast of the United States, a different version was used showing Uncle Sam with two fingers across his lips, meaning stay silent, and the printed words "loose lips sink ships." To save the cost of hiring a model, he looked in a mirror and used his own face.

Near the home of President Franklin Delano Roosevelt at Hyde Park, New York, which became a national historic site, is a museum across from the house. I purchased a copy of the "loose lips sink ships" poster there as art for our home. I chose that poster because of family history.

Before the United States was attacked at Pearl Harbor, German submarines were sinking ships from our Gulf of Mexico ports rounding the Florida coast and oil tar was washing up on the beaches. The volunteer pilots of the civil air patrol were spotting submarines against the shallow water sandy bottom. The Coast Guard was dropping depth charges on the submarines.

My father had a concession from the city of Hollywood, Florida, to take photographs across from the paved walk along the beach. He had rented a small storage room from the city, near the beach, where he stored his photographic equipment and supplies at night. Next to the storage room was a small souvenir store selling picture postcards, sea shells, and such, doing almost zero business. The couple operating the souvenir store was sitting on beach chairs. My parents struck up friendly conversations with them. They had strong German accents. The husband was frequently looking out to sea with binoculars. Both my parents had lived in apartments with German neighbors and understood German but spoke to them only in English. In these conversations my parents learned that their family had an estate in East Prussia and that in the summer he was in business in Long Branch, New Jersey, near the entrance to the Port of New York. My father went to an FBI office, probably in Miami, about 20 miles away, and reported him. Later, after the United States was in World War II, we read in a newspaper that a German spy had been arrested at Long Branch, radioing ship movements to German submarines. He was charged with being a spy, convicted, and executed.

Following the example of my dad, I have been an activist.

CHAPTER 10

∞

Evolution

The human embryo repeats the steps in human evolution. Early in the development of the embryo the beginning of gills are formed, and then deleted, showing we had a common ancestor with fish.

At a later time in embryo development we had a tail, making our embryos look like other animals with tails. In the human embryo the tail was almost always absorbed, but very rarely it is not completely absorbed, in which case the tail stub is removed surgically after birth, waiting about eight days, as with a circumcision, for the infant's blood to clot.

Brilliant evolution versus intelligent design

Look at your bare feet. Then look at your hands and compare them. Your toes resemble the fingers on your hands—why?

About two million years ago the creatures in Africa we are descended from lived near the top of trees where no big cats could climb to eat them. They had small bodies to be light weight and break fewer decaying branches. Their hands and feet were about the same; they could both grasp branches to avoid falling to the ground and being killed by the fall. Today our big thumbs resemble our big toes which are useful for walking. Our other toes are close to useless and have shortened over many generations. The way we know much about this is that one of our ancestors stepped into mud and left a small footprint that was quickly covered by volcanic lava. The present age of the lava was determined by the relative amounts of atoms in radioactive decay like uranium decaying into radium and finally into non-radioactive lead.

Today maybe Charles Darwin could have understood evolution without leaving his home by looking in a mirror at the nipples on his chest and at his bare feet and comparing them with his hands.

With intelligent design there would be no male breasts. Female breasts would be absorbed after menopause when they are no longer needed. There would be far less breast cancer.

Brilliant design allowed for evolution of nerves and networks of nerves of increasing complexity so that life could become aware of itself; aware life could also mentor its offspring, as in teaching its children how to act to survive.

After Charles Darwin published his book *The Origin of Species* in 1859, a bishop challenged him as to how the eye could have evolved. Darwin, the son of a minister, was very disturbed by what he had witnessed and the conclusions he had drawn and described in his book. Far later, the British author and evolutionary scientist, Richard Dawkins, wrote his book *The Blind Watchmaker* answering this challenge. Dawkins stated the eye had evolved at least 40 times, from the simple eye on the tail of an earthworm that detects when its tail is still exposed to light and it is in danger of being gobbled up by a bird, to the more complex imaging eyes of fish and their descendants. (For further reading, I also recommend his other books on evolution such as *The Greatest Show on Earth* and books by Stephen Jay Gould.)

With what I call brilliant design, by contrast, evolution proceeds in steps using what has evolved before. The reason orthodontists have business is that the genes defining the shape of the jaw and of the teeth are on different chromosomes. Genes defining the upper and lower teeth often are mismatched; thus the bite is deformed. This can require pulling teeth as childhood teeth are replaced by adult teeth and retainers to control the positioning of adult teeth as they grow in. My own molars grew in at an angle, impacting other teeth which had come in earlier. When I was about 20, an old dentist yanked on two molars. I remember seeing him flying across the room, almost falling down, as a tooth finally came loose. In my late 80s, I became aware that my lower molars remained; they showed up on x-rays having come in below other teeth without impacting them.

Another example of brilliant design is the almost useless male breast, in some cases harmful, as when the fat in an obese man produces estrogen leading to breast cancer. But the genetics for the male breast is along for the ride; the same very necessary genetics as for the female breast.

Humans share a common ancestor with apes. Our brain is the principle difference between humans and other apes. Our brain is our computer—it has "software." There is simple basic software for reflexes and for jumping out of the way when a tree is falling and about to hit us. More complex thoughts, as in filling out a tax return or writing this book, depend on language which has grown and evolved with us.

After climate change in Africa had reduced rainfall, grasslands were created between patches of trees which forced early man to walk. As a tree climber, man had a thumb and could hold and throw stones, create and use tools, and use weapons. Our thumb is one of the keys to human evolution. The porpoise has a larger brain but is trapped in a body without hands. Humans broke stones to create sharp edges with which they sharpened straight branches forming simple spears. Man became a predator roughly two million years ago as determined from the half-life of volcanic ash covering a footprint in mud, very much like a modern human footprint but smaller. The human body converts 35% of the calories in our food into the energy of motion, the rest into heat. (A diesel engine likewise converts 35% of the energy in the fuel it burns into energy of motion, the rest into heat to be disposed of by the vehicle's radiator.) Naked human hunters chased down fur covered animals until the animals dropped from heat exhaustion and were killed. At about the same time, two million years ago, humans began using fire to cook meat, making it more easily digestible. With more protein for their families the human population could increase.

All life on earth is related. We have microbes living within our bodies, some helpful, such as bacteria and yeast which aid in digesting food and in fighting harmful germs. There are usually far more microbes in our bodies than the far larger human cells. I speculate there may be some that act like our white blood cells, killing and digesting harmful microbes, and of course some harmful as in many diseases.

According to studies sponsored by the National Institutes of Health (NIH) or performed directly by NIH, the number of microbes living in the human body is about ten times the number of human body cells but being far smaller than human cells. They form one to three percent of the body's weight. As an infant leaves its mother's body through the birth canal it is being coated with microbes, mostly helpful, which will become part of its body.

Antibiotics are often very harmful when they are prescribed for a disease caused by viruses, against which they are useless, and kill off bacteria which are protecting us. After taking an antibiotic, such as if you had a hip or knee replacement, eat yogurt with multiple live cultures or probiotics to restore the helpful microbes within your digestive system so the harmful ones won't take over.

We are even related to earthworms. Some scientists have been studying them. To try to understand the jigsaw puzzle of genetics, great effort has been expended in the study of the very simple roundworm, C. elegans, about as simple a multicellular organism as has been found. It is a roundworm about one millimeter long found in soil. It has DNA, RNA, chromosomes and what resembles mitochondria. When mature it has a fixed number of body cells, precisely 959 body cells in the hermaphrodite type which has both male and female sex organs and can have sex with itself. It also has germ cells. The mature male-only body type has precisely 1031 body cells plus sperm cells.

Both body types produce melanin as found in human nerve cells and skin. Some human genes injected into this roundworm can replace the corresponding genes of the roundworm and function. Growth is by growth of cells, not by increasing the number of cells. This is as in humans after birth. Aging in C. elegans is studied to see the effect of insulin and other substances on aging—for hints about human aging.

Evolution is pragmatic; what works and reproduces survives. The path of evolution uses what it already has, finding additional uses. As an example, sharks evolved differently from our fish ancestors. Sharks use for their skin the same material as for their teeth.

The common house fly is an example of brilliant/pragmatic evolution. Its taste buds are on its feet so it can tell whether what it has landed on is good for it to eat. Most insects have two pairs of wings, one behind the other. The house fly has evolved its front pair of wings to be like vibrating tuning forks so that when it maneuvers rapidly, nerves at the bases of the tuning forks can sense acceleration, making a house fly hard to catch. The fly has a compound eye composed of a great many individual eyes. When its nervous system sees that the image has shifted from one eye to another it alerts the fly to danger; that a predator may be about to attack it. One strategy to squish a fly with a tissue is to approach it very slowly with little changing of what is in its field of view and then suddenly squish it. A fly swatter also works but not well if the fly is in a corner. Each individual eye has evolved with an antenna tuned to the wavelength of light, like a micro-miniaturized TV antenna tuned to the much longer wavelength of a TV broadcast.

The eye of a mole, living underground eating roots, has deteriorated—use it or lose it.

Both hearing and balance are extensions of the sense of touch. By a mutation, the ears on each side of the head duplicated themselves. The nerves for balance and for hearing are twisted together, then connected to the brain. Balance and hearing became specialized. The outer layer of the embryo before birth became skin and nerves connected to skin. We can tell whether a sound is coming from our left or our right by the tiny fraction of a second difference in the sound arrival time at our left ear and our right. The reason for the peculiar shape of the human ear is attributed to our need to distinguish between sounds arriving at the same time from in front of us and behind. Thus our ears do not have to twist around like a dog's to determine from which direction sound is coming. The sensitivity of the human ear is set so we do not hear individual molecules bouncing around, impacting our ears. As we become older, we start to hear noises in our ears that do not exist, but result from new connections from our ears to our brains to replace old connections that no longer function.

The front of a fish eye is flat. Our eyeballs are rounded. Why this difference from our fish ancestors? The fluid in the eye of a fish is like water. Fish have a mechanism within their eyes for moving dense, transparent material closer or farther from the retina so they can focus at closer or farther distances. When our ancestors came out of the water onto dry land there was a difference between the velocity of light in air and that in the fluid inside the eyes, about one third slower inside. So the eye shape changed from flat front to spherical which made the eye shape a lens. The focusing mechanism inside the fish eye, no longer needed, disappeared—use it or lose it. Muscles in the eyes we are born with changed the shape of the transparent lenses to focus at different distances, near or far. If, later in life, we have cataracts and plastic lens implants, these are fixed focus. I had cataract surgery and use two pairs of eyeglasses, one for near, as for reading and within a room, the other for distance as when outside or watching television.

We evolved in an environment with bugs that would like to make a warm nutritious home in our ears; ear hairs and ear wax are two lines of defense against bugs. Ear wax uses the same fat as breast milk. There are two different types of fat in breast milk, one predominant in China, Korea, and Japan, the other predominant in Europeans and their descendants around the world.

Tails serve two main purposes. One is to swish bugs away and the other is to help balance when running.

Sharks also give birth to tiny killing machines with the instinct to kill and eat. In the shark's mother's womb sometimes there is more than one embryo and before birth the one more developed kills and eats its twin.

The genetic code is complex. In some ways it resembles a multidimensional jigsaw or crossword puzzle. A given gene can have more than one function. Red hair was likely linked to a strong immune system. Many Irish have red hair. They are likely to have strong immune systems with the red hair "along for the ride."

Naturally blonde hair, blue eyes, and light skin appear to be genetically linked. Living in climates like northern Europe there is a great genetic advantage in having light skin to better absorb ultraviolet light and produce more vitamin D, thus avoiding rickets. Having strong bones is a great evolutionary advantage. Blonde hair and blue eyes are "along for the ride." In these days of vitamin D pills this evolutionary advantage has largely disappeared.

The vitamin D3 we gain by exposure to sunlight is not the final stage in producing the vitamin D our body uses. Vitamin D3 is converted in the kidneys to what we use. Patients on dialysis suffer from vitamin D deficiency. Taking extra vitamin D3 should help them.

The outer layer of the embryo from which we were formed grows into both our skin and our nervous system. Melanin is the brownish-black pigment in our skin and pigments in our eyes and hair that give them color. Likewise the melanin is essential for nerve function. It also protects folate from destruction by free radicals. The length of melanin particles in dark skin people is tuned to the wavelength of light to absorb light. Light skinned people have about the same amount of melanin in their skin but it is in shorter particles that let more light through to help produce vitamin D3 for strong bones. Skin color is a compromise between reducing skin cancer and having weaker bones.

There are six levels, called alleles, of melanin quantity in humans. A zero amount of melanin results in an albino with unpigmented skin and pink eyes because the capillaries in the eyes are not masked by pigments. There is an albino whale in Herman Melville's masterpiece novel, *Moby Dick*.

You may wonder how we are descended from fish, how fish came out of water, and what fish descended from. Protozoa are single-cell animals. In the sea, they are predators eating single-cell blue-green algae plants that take their energy from sunlight. Protozoa evolved to cooperate with each other, forming colonies on surfaces in shallow seas. They formed tubes that they pumped through. They ate the algae in the water. They absorbed minerals from the sea and when they died their skeletons formed what we call coral. Other protozoa colonies attached themselves to the coral, growing upward toward the sea surface. To spread their colonies, the protozoa colonies released groups of protozoa which by pumping action propelled themselves to new locations. These groups of protozoa, over time, evolved into fish. They reproduced with each other, evolved eyes, hearing, and fins to propel themselves better. They became fish. Some fish swallowed air to inflate themselves, to be harder to swallow and, at low tide, took oxygen out of the air while still using their gills to take oxygen out of water when the tide came in. In an aquarium, I have seen creatures with long, spindly legs for crawling along on the sea floor. We are descended from fish like all other land animals.

We, like other descendants of fish, inherited the split brain of fish, which contributes to our capacity to have second thoughts, inhibitions—before actions. In fish, the purpose of the split brain, two hemispheres, was to allow one hemisphere to sleep while the other hemisphere was awake and controlled swimming so that the fish could absorb oxygen into their gills from the water the fish was swimming through. Investigators noted that fish swim in circles, first in one direction, then the other according to which hemisphere is awake. Our ancestors, high in trees, revived and strengthened this capability. This was especially important to females nursing infants and young children. It allowed them to sleep in treetops while holding on during sleep without waking up and falling. Have you ever had a dream that you were about to fall?

An exception to this sleep pattern are flatfish which lay on the sea floor, their skin like a chameleon's changing appearance to match the surrounding sea floor. Flatfish such as flounder and sole are born with eyes on both sides. The eye on one side gradually migrates around so that both eyes are on the same side. The bottom of a flatfish is white so when it leaves the sea floor and is swimming above other fish it more closely blends in with the sky above the sea. Flatfish evolved independently twice; once with the eye migrating to the right side, once to the left.

Why do we have bladders to hold our urine until nearly full and then suddenly release it? Why not just continuously release it drop by drop as it comes out of our kidneys? It is not because it would be messy and offensive in polite society. It is because a bladder has survival advantage. Otherwise predators could locate a good meal by smelling the trail of urine. Likewise, solid waste is expelled all at once and fish and their descendants move rapidly away from their smelly waste.

Our bodies evolved to release salt. Life evolved in sea that was much less salty than today. Rain leached minerals out of land and flowed into the seas; more entered from the sea floor.

"The Rime of the Ancient Mariner" includes a line, "Water, water everywhere but not a drop to drink." The blood of fish and their descendants including us is less salty than the seas. Sea birds drink sea water and get the excess salt out of their bodies through their running noses.

Lightning is of extreme importance to the evolution and existence of life. Lightning between clouds of opposite charge, and lightning between charged clouds and the earth's surface, provides the extreme heat energy which combines the about four-fifths of the atmosphere (which is nitrogen) with the almost one-fifth that is oxygen. Precipitation washes the combinations out of the atmosphere to the earth's surface. This fertilizes the land and water, providing the nitrogen for the protein in our muscles and the nitrogen compounds essential to heredity—DNA, RNA, etc.

Some butterflies evolved with patterns on their wings that look like the eyes of a large bird. When a predator suddenly sees them it is frightened away. Some lizards are colored to look like the bark of trees they rest on, to be more difficult for a hungry predator to find. A chameleon is a lizard that changes colors and patterns to look like what it is resting on. Some frogs are deliberately conspicuous; bright yellow, red, or green. They have poisons in their skin that sicken predators so the frog's siblings and other relatives will be avoided by the predators that ate them.

Some lizards have a different main purpose for their tails. It is to help them escape predators chasing them. They pinch off their own tails and leave them wiggling on the ground. The predators eat the tails and the lizards escape and grow replacement tails. This ability to replace a part of the body is of great interest to researchers. Replacing or repairing the hippocampus near the juncture of the two hemispheres of the brain would allow older people with dementia to convert short term memories into long term so they would not be always living in the past. Repairing brains that had undergone traumatic injuries by stimulating stem cells with electric currents, and replacing other organs (liver, pancreas, etc...), and perhaps eventually restoring vision to the blind by re-growing eye retinas (which are extensions of the brain) would win researchers well deserved Nobel Prizes.

The purpose of procreation is to compensate for and take advantage of mutations which are advantageous, often in a changing environment.

There are many different types of imaging eyes as well as simple non-imaging light sensors on the tails of earthworms. Their function is to inform the earthworm, that may have come to the surface when the ground was moist and burrowed back into the ground, that it is fully underground so a bird cannot find and eat it.

The eyes of fish and their descendants, including us, have a blind spot at the center of the retina called the fovea. The nerves from the retina to the brain start at light sensing cells at the front of the retina and pass through the fovea to the back of the brain where the images are processed. We compensate for the blind spot by unconsciously vibrating our eyes to see what would otherwise be hidden by the fovea blind spot.

The octopus and its relatives are descended from mollusks. They also evolved eyes with pupils. Their eyes differ from ours. The nerves that connect their light sensing retinas to their brains are behind the retina and they have no fovea blind spot.

We have the same atoms and thus chemistry throughout the universe. We know this because the light from stars has spectral lines (color bands) as we find here on earth but shifted in color, mostly more reddish, according to our relative motion with respect to those stars. This color shift is known as the Doppler effect, the same as the changing tone from higher to lower pitch as the whistle on a train passes us. The Doppler effect is used by police to measure the speed of vehicles on highways. We have the same biochemistry as life elsewhere in the universe.

There is a difference (that seems very peculiar) between turning around in one direction and turning around in the other because of something in quantum mechanics called "spin." You have to turn around twice to be back where you started from. I believe the direction (clockwise or counterclockwise) of the spiral twist of the genetic code in our chromosomes is the same throughout the universe.

The purpose of blood is to bring oxygen to the cells throughout our bodies from gills or lungs. The oxygen combines with blood glucose to provide energy for motion and to warm our bodies. On earth and elsewhere there are two types of blood. One is based on iron, as in fish blood and our blood. The other is based on copper, the blood of crabs, horseshoe crabs, mollusks and their descendants (like the octopus). Insects and spiders do not have blood but have tiny tubes that bring air directly to their tissues.

Within our blood, the iron in the hemoglobin is not static. The iron goes between two states. "Hemo" means iron. The iron in hemoglobin releases oxygen by going from one state to the other. The two states are called ferric and ferrous (ferric is one oxygen atom to one iron atom, ferrous is three oxygen atoms to two iron atoms). Similarly, cupric is one copper atom to one oxygen atom and cuprous is two copper atoms to one oxygen atom. Carbon monoxide latches tightly onto hemoglobin and prevents it from functioning. When this happens, our blood can no longer hold and transport oxygen.

Life on earth depends upon chlorophyll in blue-green algae in the sea and chloroplasts in land plants to absorb sunlight, and carbon dioxide from the sea and atmosphere to produce starch, cellulose, and sugar. We are among the predators who attack plants for our food and fibers for our clothing.

Chlorophyll in plants absorbs energy from light and helps plants grow. Chlorophyll contains copper. There are two types of chlorophyll; both absorb energy from light in a process involving two photons. A photon is both a wave and a particle. The type in annual plants is more efficient but the plant is killed by frost. The other type, in perennial plants, is frost resistant but less efficient. Crabgrass is an annual that regrows each year from seeds produced the year before. Because crabgrass has more efficient chlorophyll than regular perennial grass it can often crowd out the regular grass, leaving areas of brown, dead crabgrass in the winter. The color we see in plants, mostly green, is the light reflected and not used for photosynthesis.

Just as genes from mitochondria have mitochondria in symbiosis, some DNA and other heredity material has migrated into mitochondria of plants in what is called promiscuity (the same word as applied to humans). It has evolved to help the plants function.

Evolution is always a work in progress, adapting to a changing environment. In apes, males impregnate females with sperm from behind as other mammals do. In human evolution the female vagina has gradually migrated toward the front. Unlike a female dog which has multiple nipples for nursing its infants but does not have enlarged breasts, human female breasts are enlarged. This has been attributed to their resemblance to the female buttocks with the purpose of encouraging human males to have sex from the front.

Among the ways humans differ from other apes is that of the skull after birth. In humans the bones that form the skull do not fully join together until after the infant's brain has had months to grow. In apes this happens soon after birth.

The human diet has changed since the beginning of domestication of animals. A gene allows an infant to digest its mother's milk. Milk contains a sugar called lactose. A genetic change induced by mutation allows the child growing into an adult to continue digesting dairy products. Some individuals do not have the capability to digest lactose. Instead, the lactose in the digestive system, aided by yeast there, ferments, producing gas—then you are a brewery. There are pills and liquid to remedy this by helping to digest lactose.

Mucous has the important function to protect our digestive systems from the strong acids our body makes and uses to digest solid foods. Without it those strong acids would also digest our digestive tract. Every creature that eats solid food makes and uses mucous to protect its digestive system—so this originated very early in evolution.

Contrary to common liberal belief, and what is often taught, heredity predominates over environment which includes education and those we associate with. Family is both heredity and environment. The question often asked, as by former president Obama during his campaign to become president, while addressing the troubles of the native born African-American community was, "Where are the fathers?" He should have asked, instead or first, "Who are the fathers?"

A rat, by instinct, will kill the offspring of a rival rat as well as that rival and then make the "widow" pregnant. Another example is the wild dogs of east Africa. The leaders of their packs are females. The leader kills the litter of another female dog in her pack, except for sparing one which keeps her nursing. She uses her as a "cow" to nurse her own litter. This behavior is instinctive, hereditary.

A human baby is born immature relative to other species because of its large head (with brain) passing through the birth canal. The baby instinctively suckles on its mother's breasts, otherwise it would starve. A marsupial infant is born very tiny and immature. It has strong front legs with which it climbs from the birth canal up into a protective pouch where it nurses on its mother's nipples. Marsupials include kangaroos and, in North America, possums.

The Tasmanian devil is a marsupial that has evolved on the island of Tasmania off the southeast coast of Australia. It is a wolf-like predator whose teeth have evolved to be very much like that of a wolf. It illustrates how environmental pressures can force very different creatures to evolve to resemble each other. Form evolves to follow function. Whales and porpoises are descendants of land mammals that have returned to the sea. Likewise, the tails of whales and porpoises resemble those of fish but are moved up and down like the feet of a human swimmer rather than from side to side like a fish. Genetic testing has revealed the whale is descended from the hog-like hippopotamus swimming in the rivers of west Africa. The need for efficiency of propulsion was the driving evolutionary force that caused the same fin shape to develop for very different species. Similar fins will have evolved in similar seas in other solar systems.

Fur seals and harbor seals, both of many species, live in mostly cold waters around the world. They appear to be descended from two different land animals that went back to the sea. Fur seals propel themselves through the water primarily using their "forearms." Harbor seals propel themselves primarily using their "rear feet." This illustrates similar but not identical behavior along very different evolutionary paths.

Human behavior likewise responds to environmental pressure, not always in ways we would consider good. Consider rape. Over thousands of years a male who raped a female was likely to have more descendants than one who did not. The female who was raped, or promiscuous, was also likely to have more descendants. These descendants are more likely to have inherited this behavior tendency. "Like father, like son."

Compare the turkey vulture which must survive in a wild environment with a domesticated turkey which has its food given to it. Over many generations, the domesticated turkey becomes stupider and stupider; hence the insult comparing someone to a turkey.

Consider bright professionals who delay having children and have fewer children because they spent longer periods of time in educational and career settings. The average intelligence in the world is thus declining—evolving backwards.

To avoid becoming extinct a creature in the wild must reproduce before it is eaten. There is greater pressure on small animals like mice than on larger creatures that can better defend themselves. Mice have short pregnancies. Because on earth we have had symbioses (the merging of species) and mitochondria tend to be produced from precursors more slowly than cell nuclei, thus aging is far more rapid in a mouse than in larger creatures that can afford to take the time for a longer pregnancy.

Another example of evolutionary pressure to reproduce while still small is the tuna fish. Tuna are caught in huge nets towed by a pair of boats. There are openings in the nets large enough for immature tuna to escape through. Some very small bodied tuna, mature enough to reproduce, also escape; tuna will evolve to have smaller bodies.

Human females generally have smaller bodies than males and thus need less food. But the difference is much less than for other apes. With smaller bodies than males, females burn fewer calories than males who are "burning candles at both ends." Females age more slowly and in the United States, now that death during childbirth is rare, about 70% of those alive at 100 are female. Within other species, using dogs as an example, small bodied dogs live much longer than big bodied breeds.

With smaller bodies than males, human females have lighter weight brains. They likely have about the same number of brain cells but smaller because of what is called cell constancy within a species. The number of nerve cells a female brain nerve cell connects to is greater than for a male, helping compensate for a lighter weight brain.

We have evolved together with other life on earth. This evolution includes microbes that help us in digesting food, especially fats. Helping our immune system fight harmful microbes is essential otherwise we could not survive.

To have a civilization, individuals must cooperate. This requires language and the inheriting of a feeling for family, clan, and nation—with empathy—and a willingness for sacrificing themselves if necessary to protect their family, clan, and nation.

In contrast to humans, bears are solitary creatures except for female bears with cubs who in cold weather live together in dens. Males have their own solitary dens. Bears would not evolve into the basis for a civilization as humans did because they are not social animals, lacking the ability to communicate with complex language and lacking hands capable of handling tools or even picking up a stone and throwing it.

On earth, humans are unique. It is probable that in the vast majority of solar systems with creatures, very few have evolved with the capabilities to create a civilization. I base this statement on the fact that none did, after meteorite/comet impacts on earth, until the one about 66 million years ago that eliminated dinosaurs (except birds) and allowed tree climbing creatures (thus capable of throwing stones, sharpening stones, sharpening sticks to use as weapons, then attaching sharpened stones to sticks to use as spears and arrows). Meanwhile, being social animals, they evolved language and the brain capacity to use it.

Guided by biology on earth, we can intelligently speculate about the biology of space aliens, whether here or not, from biochemistry and the paths of evolution here on earth.

"We cannot see the forest because it is hidden by the trees." There are an estimated 150 billion stars, not including another 100 billion or so brown dwarfs, in our galaxy. It is difficult to observe these stars from within our own galaxy. We can, however, see the relatively near Andromeda galaxy. Although it is larger, it is nearly a twin. We view it from the side. Our galaxy is roughly 100,000 light years across. The Andromeda galaxy is about 2.5 million light years from earth. Astronomers using telescopes do the equivalent of looking through a pinhole to count the galaxies in space figuring how the view through the pinhole compares with looking in all directions. Then they count the galaxies seen through the pinhole which is something like 20,000 depending on the size of the pinhole and the distance to the pinhole. The number of galaxies is about the same looking in any direction in the universe. Likewise, they can look at the Andromeda galaxy through the pinhole, count stars and estimate how many are in the whole galaxy, thus estimating 100 million to 150 million.

There are roughly 150 million galaxies in the observable universe. Even if only one in a million stars in the universe have planets or large moons with the right conditions to support life and evolve civilizations, there are more civilizations in the observable universe than humans who are now alive or have ever lived on earth.

What are the conditions for civilizations to evolve in a solar system? There must be a planet or a large moon at the right distance from its star/sun for liquid water to exist. The planet or moon must have enough mass and gravity to retain an atmosphere and oceans, keeping them from disappearing into space. As earth formed, it was too hot to retain water or an atmosphere. Mars, much smaller than earth, has canyons on its surface which seem obviously to have been cut by water. It is likely but not yet fully proven that life once existed on Mars. Some minerals like carbonates that are produced by life have been identified by equipment we have landed on Mars. At the north and south poles of Mars there are changes in appearance with the seasons of the Martian year. These have been interpreted as melting ice, usually covered by dust.

If the planet or moon is too massive, its accumulated oceans will completely cover its mountain peaks. A requirement for a civilization to evolve is that the balance be just right between the planet accumulating water from space and losing water to space—so that both oceans and land exist. This is called the "Goldilocks" situation.

Our solar system is fortunate in having the planet Jupiter. Because of its huge mass, it attracts comets from outer space which would otherwise endanger us. The astronomers Shoemaker and Levy spotted a comet coming into our solar system and as it approached Jupiter it was disrupted by Jupiter's gravity into 21 pieces large enough to be identified; one piece crashed into the gas giant Jupiter, created a dark area as large as the earth. In some other solar systems, a massive planet may have a close orbit to its sun. Then it will be much more effective in capturing comets than Jupiter is.

Earth is in a cosmic shooting gallery with comets, Apollo asteroids (those with orbits that cross the earth's), and very rarely an object that passes through the plane of our solar system, (and perhaps our whole galaxy) at an angle. At random intervals of about 100 million years an object like that which killed off the dinosaurs crashes into the earth creating punctuated evolution.

During most of the about four billion years the earth has existed the atmosphere has not been right to support the evolution of human life. The early atmosphere was probably mostly methane, the main ingredient in natural gas. This would have prevented the oxygen in the air we are now breathing from accumulating. Oxygen would have reacted with methane to form water, carbon dioxide, and carbon monoxide. There is evidence that life as we know it began 600-some million years ago following about a 100 million year period when the land and the oceans were covered in "snowball earth." Carbon dioxide from undersea volcanos finally accumulated enough to break through the ice into the atmosphere and slowly change the climate so the ice could melt. The sun is much warmer now than it was when "snowball earth" started, possibly from the sun having a very long period of the sun's surface being a little cooler. Life survived at undersea gas emitting vents (which still exist).

The vent gas supported a very different type of metabolism than we have today. Now breathing is oxygen based.

With genetic engineering, there is great danger that human heredity will be changed in future generations by attempts to treat diseases; "they know not what they do." The damage may not be apparent until double doses of recessive genes appear in human grandchildren and later future generations; then the damage can never be undone.

Evolution applies to much more than biology. Using what you already have also applies to language—English today is far different than in the time of Chaucer, or even a half-century ago. Consider all the new words of technology and the changing meaning of existing words.

Benjamin Franklin, one of America's founding fathers and inventor of the lightning rod and a more efficient wood burning stove, wore a chain attached to eyeglasses. Later, another use for ears was found: to hold eyeglass frames in place to help vision.

As another example, consider the improved steam engine of James Watt developed in Great Britain around the time of the American Revolution and first applied to pumping water out of coal mines instead of having horses walking around in large circles harnessed to turn a pump shaft. Thus the term horsepower, first used to advertise Watt's steam engine.

The steam engine was soon applied to railroad trains and boats. The first trains were in Great Britain. The distance between the rails matched the ruts made in the ground by Roman chariots in the centuries when Britain was part of the Roman empire. Steam engines replaced horses used to pull carts a short distance to docks. Rails should have been spaced farther apart for better train stability.

In some places, oil was seeping from the ground but was difficult to collect. In 1859, just before the start of America's Civil War, a pipe was drilled and forced 69 feet into the ground at an oil seepage site in northwestern Pennsylvania. Overnight the iron pipe filled with oil and the world's oil industry was born.

By distilling oil, as with whisky, the oil could be separated into kerosene, valuable for lanterns and much less expensive than whale oil, helping save whales from extinction. The left over by-product was very inexpensive. Therefore, the steam engine's back and forth motion of a piston-in-a-tube was adapted to become the internal combustion engine and the cheap by-product called gasoline was used as fuel. A later variant was the diesel engine, much later, the gas turbine. Without the piston-in-a-tube steam engine coming first what would the history of the internal combustion engine and motor vehicles have been?

CHAPTER 11

∞

Nuclear Weapons and Einstein

It is important to make people around the world aware of the actual danger of nuclear weapons. One should not be misled by the relatively pipsqueak (tiny) detonations of the atomic bombs that ended World War II—or be oblivious before they are injured or killed in a "nuclear 9-11." Hopefully, informed people may be able to prevent disaster.

First, for context, the cast of characters and the chain of events. Cast of characters: Albert Einstein; his friend Leo Szilard; British science fiction writer H. G. Wells; Lisa Meitner; American President Franklin Delano Roosevelt; physicist, Nobel Prize winner Enrico Fermi; physicist, spy Klaus Fuchs; Chinese dictator Mao Tse Tung; Winston Churchill; Joseph Stalin; President Harry Truman; Nikita Khrushchev; physicist, nuclear weapons designer, Nobel Peace Prize winner Andrei Sakharov; Ayatollah Ruhollah Khomeini; and his successor Ayatollah Ali Khamenei.

When the Hiroshima bomb exploded in August 1945 I had already taken an undergraduate course in atomic and nuclear physics and knew that nuclear weapons were possible. I have been closely following what has been published ever since. I have also worked on delivery of nuclear weapons by both missiles and aircraft, and defenses against them, including the United States' first generation intercontinental ballistic missiles (ICBMs). For credibility, I mention that as the ICBM program was about to start, for obvious reason, IBM asked the department of defense to capitalize the c in the middle of the word intercontinental.

I have had American security clearances including "top secret with special-need-to-know," but I have never had the special clearance to know the inner workings of nuclear weapons. What I have written here can be found on the internet by nuclear weapons designers anywhere in the world, but is not understood by politicians who, with a few exceptions, have no understanding of science. Smuggled nukes can be set off by timers or cell phones with the culprits out of the country before detonation. Smuggling nukes is not new. During the Cold War, nuclear demolition charges were smuggled into NATO countries, including the United States, by the KGB for the purpose of disrupting government and military command and control, creating chaos should the Cold War turn hot. In the United States they were reported to have been stockpiled in Virginia convenient to targets in the Washington, DC area. As tactical nuclear weapons, which also include nuclear artillery shells, there is no "pushing a button" by the head of state necessary to activate them. For the stockpile in the United States, trusted KGB agents could activate and detonate them. Tactical nuclear weapons have as little as 50 tons of TNT equivalent, like the American Davy Crockett rocket of the 1950s, to about 1,000 to 1,500 tons. I believe when the Cold War was interrupted, Gorbachev ordered the nuclear demolition charges to be smuggled back out.

It is important for you to realize that the atomic bombs that were ready to end World War II were only the inefficient fuses designed to set off thermonuclear bombs (the so-called hydrogen bombs). Bombs with a detonation energy a dozen times the Hiroshima bomb could be designed for smuggling and be small enough and light enough to be carried in any car.

On the internet, type in "W58 missile warhead" and see what comes up. The W58 was the warhead on the Polaris A-3 missile, three per missile for a shotgun effect on a target area. The W58 warhead was rated at 200,000 tons of TNT equivalent. A version of the W58 designed for smuggling would delete heavy shielding of the bomb from the heat of air friction on reentering the earth's atmosphere at extreme velocity like five miles (eight kilometers) per second. It would be lighter and shorter, able to fit in any car, placed there by two people. Heavier bombs with five times the energy (one megaton of TNT equivalent) could be carried in an SUV or pick-up truck but would be more difficult to smuggle.

Also type in "Tsar Bomba" which means "king of bombs." The Hiroshima bomb was the only one of its type ever detonated and from damage done, including baking clayish soil 2,500 feet below it into the consistency of pottery, the energy released was estimated to be equivalent to 12 to 15 thousand tons (kilotons) of TNT, later raised to 16 kilotons. By contrast, the Tsar Bomba detonated over an unpopulated island in the Arctic Ocean was redesigned to reduce weight and environmental damage such as blowing some of the earth's atmosphere into outer space and reducing radioactive fallout. The Tupolev turboprop bomber that would drop it could take off and climb to about 45,000 feet altitude in order to drop it with a parachute and give the plane time to get far enough away to survive the detonation. It was detonated in October 1951. The shock wave followed the earth's curvature since the velocity of sound and shock waves is faster in warmer air near the ground than colder air at altitude. It broke windows hundreds of miles away. Its objective was to intimidate President Kennedy the year before the Cuban missile crisis. The bomb was thermonuclear, a so-called hydrogen bomb. Much of the energy came from plentiful lithium, which is in rechargeable batteries. The energy release was over 4,700 times the Hiroshima bomb.

Mass is converted to energy. Energy can be defined according to the mass that disappears upon impact. The energy of a bomb can be defined according to mass of the atoms within the bomb which disappear upon detonation. This mass has been converted to energy. The mass, amount of matter, of the Hiroshima bomb converted to energy is estimated at 700 milligrams. This is about the mass of a dime. In comparison, the meteorite that eliminated the dinosaurs is estimated to have had about a million times the energy of the Hiroshima bomb. Compare your body mass to this. It does not mean that you can be that destructive.

The time for an ICBM to go either way between the United States and targets in Russia, China, and Iran is less than it takes me in the morning to shave, wash up, get dressed and sit down for breakfast, roughly half an hour.

The United States has had two false alarms that I am aware of. One was caused by a very realistic training tape being placed in the system of the North-American Radar Air Defense (NORAD), then underground at Cheyenne Mountain, Colorado Springs. Not everyone there was informed it was a training tape and bombers with nuclear armed stand-off missiles took off. At that time, we had flight crews in the bombers 24/7 ready for take-off on their way to the Soviet Union. Because it would have taken too long to get from barracks to planes, the bombers could be recalled when the time had passed for ICBMs to have arrived. The bomber bases were across the country near our border with Canada.

The other false alarm was soon after the Ballistic Missile Early Warning System (BMEWS) had been activated. The main part of BMEWS was in Thule, in northern Greenland. It had a football-field-size radar antenna standing on a cliff overlooking the sea. When Canadian geese flew closely in front of the antenna they fell down dead, cooked as in a microwave oven. A warning that a massive ICBM attack had been detected was reported by the system. It was a false alarm, traced back to the moon having come into the radar beam and reflected back from many pulses before. This was remedied by having each pulse slightly different so they could be distinguished.

The Chinese and Russians, by backing Iran and Pakistan, are putting themselves at risk. China had largely put Pakistan in the nuclear business in the 1980s, to make trouble for India, giving Pakistan the design for China's first (1965) nuclear bomb which detonated with a little less energy than the Hiroshima bomb. China has built four Chernobyl type (with graphite moderators) nuclear reactors in Pakistan to produce plutonium for Pakistan's nuclear arsenal. Pakistan's nuclear weapons are controlled by the same intelligence and military people who placed a ship off the coast of India to launch inflatable boats with outboard engines which brought Muslim terrorists on a suicide mission to the coast of Bombay/Mumbai. They are not controlled by Pakistan's president.

China has built the world's greatest target, the Three Gorges Dam, on the Yangtze River. If the dam is removed by a small nuclear weapon, perhaps smuggled, the wall of water released would reduce China's population by very roughly 100 million people who live downstream, including cities like Shanghai. The leaders of China know this, but need to be reminded that the rest of the world, including those who would blow up the dam, know this also. The Chinese government can estimate the number who would be killed by the wall of water released. It will depend upon the time of day or night, where the bomb is placed behind the dam, topography, how much warning, and how fast people can get to high ground, etc.

The Former Soviet Union stupidly provided missile and nuclear technology which, like a boomerang, may come back and hit it. What actions are our government leaders taking to prevent nuclear disaster?

Albert Einstein, trying to skip his last year of high school in Germany, applied to the Polytechnic Institute in Zurich, Switzerland. He failed to pass the entrance examination. He tried again a year later and was accepted. He graduated and continued on in graduate school. In 1905 Einstein submitted his PhD thesis. It was rejected as not credible. The title was "The Special Theory of Relativity." In it he asked himself what would happen if he traveled faster and faster, approaching the velocity of light. His theory included the equivalence of mass and energy. Energy and mass can be changed into one another. In the Hiroshima bomb the mass that was converted to energy was about the mass of America's lightest weight coin, a dime. Only a fraction of one percent of the energy in the nuclear material in the vaporized bomb reacted before the reactions stopped as the vaporized bomb blew apart.

Not having the PhD credential, Einstein was not employable as a professor at the university he graduated from, or others, so to support himself he went to work as a clerk at the Swiss patent office. One of his friends was a chemist/physicist from Hungary, Leo Szilard. They received many patents together including a way to refrigerate.

The writer, H. G. Wells, became acquainted with Einstein's theory of relativity and the equivalence of mass and energy. He wrote a book about a bomb based on this and Leo Szilard, reading it, was inspired to design a bomb based on the element beryllium which is a key ingredient in the chain reactions of all nuclear weapons.

When the uranium or plutonium in a nuclear bomb fissions, among the fragments are helium nuclei called alpha particles. When alpha particles collide with beryllium, neutrons are released helping the explosive chain reaction along. For safety in storage the alpha particles are prevented from colliding with beryllium before the bomb is vaporized by plating the beryllium with a heavy metal like gold.

Leo Szilard was ridiculed in the magazine section of the *New York Times* by an ignoramus writer for stating that a nuclear weapon could be based on beryllium. With what was known at the time Szilard made that statement it was reasonably informed.

What I know about the inner workings of nuclear weapons is based on what was publicly released by governments and as a physicist "putting two and two together." I can figure out how to design a nuclear weapon, as can physicists anywhere in the world. There is also what I believe is deliberate misinformation publicly available, as on the internet.

During World War II the American government stupidly designated the Yagi antenna (later used for TV reception) as classified information even though it had been invented by Professor Yagi at the University of Tokyo. Much later, the American government classified some laser technology. It was unaware the technology had been openly published in Russian but not translated.

The chain of events that led Einstein and his associates to induce President Franklin Delano Roosevelt to consider developing nuclear weapons before the United States found itself in World War II (on December 7, 1941, with the Japanese attack on Pearl Harbor) is described in what follows. Funding to develop atomic bombs became available once the United States was at war.

Enrico Fermi received the Nobel Prize for physics in 1938 for his discovery of radioactivity induced by radiation by slow neutrons. He was allowed to bring his wife and her parents to Stockholm, Sweden, for the award ceremony which included his acceptance speech. His wife was Jewish.

They never went back to Italy but instead went to the United States.

Fermi went to the Office of Naval Research in Washington, DC, and tried to persuade them to provide funding to investigate making nuclear weapons. He was figuratively "thrown out on his ear" and called "that crazy wop," an insulting term for an Italian. The physics community was small and Fermi knew Leo Szilard who then lived in Washington and supported himself with research study contracts from the Office of Naval Research. Szilard knew the result of a study before he received the grant for it because he had already performed the study before he proposed it; thus his studies always turned out as proposed.

Fermi became a professor at the physics department of the University of Chicago. He built the world's first nuclear reactor underneath the seats of the university's sports stadium. The reactor used graphite (the so-called lead in pencils) to slow down neutrons so that they could be captured by uranium atoms causing fission (splitting). To inform the physics community on the east coast, he sent a coded message. This message to Isadore Rabi (whom I later met) at Columbia University was worded so that Professor Rabi would understand it. The message said, "The Italian navigator has landed and the natives are friendly."

Fermi died of cancer a few years later, probably caused by his exposure to radioactivity.

Einstein was on a lecture tour of the United States in 1933 when Hitler came to power. Two Jewish families were major donors to Princeton University. The Bamberger family, that had sold their chain of department stores in New Jersey to Macy's, had earlier been major funders of the Institute for Advanced Study at Princeton University. Einstein went there to live and work. Lewis L. Strauss served at sea under President Theodore Roosevelt becoming an admiral and serving in Roosevelt's cabinet. Later, he became wealthy as a businessman and helped fund, establish, and became a trustee of The Institute for Advanced Study. Strauss was also president of Temple Emanuel on Fifth Avenue in New York. The temple has a museum about its history which includes an exhibit about Admiral Strauss.

After truck bombs were detonated in Beirut, Lebanon, while Reagan was president, barriers, some disguised as large reinforced (with steel rods) concrete flower pots, were placed around the White House and many other federal buildings in Washington. Temple Emanuel likewise erected barriers, so-called "Jersey barriers" designed by engineers at the Federal Bureau of Roads because they were first installed as road dividers in New Jersey.

In 1938, Leo Szilard wrote two letters to President Roosevelt advising him that nuclear weapons were possible that could be carried on a ship into a harbor to blow up and destroy the seaport and that America should be the first to build them. He wrote two versions, one long, one short. Einstein was then on summer vacation at Cold Spring Harbor near the tip of Long Island. The reason the example of a nuclear bomb aboard a ship was used is that the weight of such a bomb was uncertain.

Szilard asked a young physicist, Edward Teller, to drive him from Washington to Einstein at Cold Spring Harbor. Teller later was among the principal designers of the so-called hydrogen bomb. Einstein chose to sign the longer version. Lewis L. Strauss had an appointment to see President Roosevelt two weeks later and handed Roosevelt the letter Einstein had signed. Roosevelt gave the letter to his chief of staff, "Pa" Watson, and asked him to get a second opinion. Watson asked Vannevar Bush, Dean of Engineering at MIT, and head of Roosevelt's Science Advisory Board. To make sure the advice from the Science Advisory Board would not be biased by Einstein's signature, Watson did not show Einstein's signature to Bush. The board agreed with

Einstein's conclusion that nuclear weapons were possible, but funding for atomic weapons did not become available until after the United States was in World War II.

Einstein and Szilard had left wing friends and therefore were under suspicion and could not get the security clearances necessary to participate in the design and development of nuclear weapons. Einstein later did some unclassified calculations for the navy.

In 1942, physicists who had been dismissed from their positions at German universities (because they were Jewish) succeeded in getting out of Germany after Hitler came to power. They got teaching positions at American universities, and on their own initiative, designed the basis for thermonuclear weapons—so-called hydrogen bombs. The fuse necessary to ignite thermonuclear weapons was the only part of the thermonuclear weapons ready for World War II. This fuse was used in the Hiroshima atomic bomb dropped on Japan, helping end World War II in August 1945. The Nagasaki bomb, dropped a few days later, was designed differently and did not use that type of fuse.

Albert Einstein was never awarded the Nobel Prize in physics for his Special Theory of Relativity (1905) or his General Theory ten years later. In developing the General Theory, he began by asking himself if the force on your feet is a different kind of force if you are in an elevator where the movement is gradually changing versus in a stationary elevator. Is there a difference between gravity and acceleration? This led to his theory that the attraction of a mass of matter distorts space and time. Carried to its conclusion, it led to the idea of black holes, where with sufficient mass in a limited volume the attraction can be strong enough so nothing, not even light, can escape.

Einstein was unsettled about some of his theories. He was skeptical of the black hole prediction based on one of his own theories. In comparison, his predictions based on his theory of General Relativity have proven to be correct. He explained why the orbit of mercury has a wobble. Newton's theory of gravity says that bodies attract each other and, therefore, the wobble of mercury's orbit would not make sense. Einstein's theory explained the wobble.

In 1921, the Nobel committee awarded Einstein the Nobel Prize for physics "for his services to theoretical physics and especially for the photoelectric effect." He explained that light was acting as both particles of energy and as waves. Making light brighter does not increase the energy of the photon particles. To increase the energy of a photon particle (radiation) you must shift it toward blue radiation waves. For example, a TV remote uses the photoelectric effect by using infrared radiation too low in energy to be seen by our eyes.

He also explained why on a clear, sunny day near noon the sky is blue and near sunrise and sunset it is red. Red light waves are about double the wavelength of blue. Therefore, they penetrate through the atmosphere better than blue. If you consider a cube of air particles one wavelength on a side, there are eight times as many particles in the cube of red wavelength as of blue so the number of particles in the red volume averages out much better than in the blue cube. The blue light we see looking up near noon was scattered, leaving the red light to go farther, toward the horizon, before it is scattered. The scattering is because air molecules bounce around so there are sometimes more, sometimes fewer, molecules in any given volume of air.

Sometimes you can see a beam of light coming through a window illuminating particles of dust bouncing around. They are bouncing because the dust particles are being hit more on one side than the other. The air molecules in a room are moving faster than a jet plane. You are not knocked over because you are being bombarded equally by air molecules from all directions; when there is a slight difference we call it wind.

In Germany, Werner Heisenberg was the leading ethnic German physicist and Hitler declared that physics was a Jewish science. That placed him in deep trouble. His mother knew the mother of Joseph Goebbels and used that influence to protect her son. Goebbels was the propaganda chief for Hitler. Goebbels explained the inconvenient fact that while Hitler idealized the image of the tall blue-eyed blond Aryan, both he and Hitler were short and dark haired. Goebbels stated there was a second type of Aryan German (a little like, in fairy tales, dwelling in the Black Forest). Goebbels stated that if you tell a lie often enough people will come to believe it. His advice is very often followed by politicians.

Heisenberg was born in 1901. He won the Nobel Prize for physics in 1932 for his work developing the uncertainty principle and helping found quantum mechanics. The uncertainty principle states that on an atomic scale when you try to measure the position or the momentum of an object you disturb it and the product of the uncertainty in position multiplied by the uncertainty in momentum is a constant called Planck's Constant. Momentum is mass times velocity. In 1900, Max Planck had arrived at his constant trying to explain heat radiation coming out of a hole in a hollow, hot object called a black body. The energy in a particle of light is called a photon. Light is both a wave and a particle.

During World War II, the Norwegian resistance sank a ferry carrying heavy water across a lake. Heavy water is made using a great deal of electricity, as from a hydroelectric power plant. It is called heavy water because the hydrogen atoms in it, called deuterium, are double the weight of other hydrogen atoms. Heavy water in a nuclear reactor can slow down neutrons so they can be captured by uranium converting them to plutonium for use in nuclear weapons (like that dropped on Nagasaki) helping end World War II. That loss of heavy water combined with a calculation by

Werner Heisenberg showing that atomic bombs would be too heavy to use ended Germany's atomic bomb work. It may have been deliberate. Heisenberg kept his mouth shut about his calculation for the rest of his life.

CHAPTER 12

∞

Principles of Mismanagement

I have knowingly participated in mismanagement to "avoid making waves" and keep my job—not to be high on the layoff list when because of funding difficulties, "a reduction in force" becomes necessary.

The recession in the United States beginning in 2006 that spread to the rest of the world was largely due to mismanagement. Managers who were cautious and reduced risks were replaced by others who showed higher profits on paper but erected a "house of cards" which collapsed suddenly.

Parkinson's Law

Cyril Northcote Parkinson was a British civil servant who observed that as the number of ships in the British Navy declined after World War II, the number of civil service employees of the navy increased. Likewise, as colonies became independent countries after World War II the number of civil service employees in the colonial office increased; it had the greatest number of employees when it was folded into the foreign office because of a lack of colonies to administer. He explained that officials want subordinates, not rivals, and officials make work for each other. He noted the number of employees in a bureaucracy rose by five to seven percent annually "irrespective of the amount of work (if any) to be done."

The Peter Principle

A book on management called *The Peter Principle* sought to explain why things seem to always go wrong. It states that people in organizations get promoted beyond the level of their competence. I disagree. Sometimes they are promoted even farther by someone who is already above the level of his competence. Also, sometimes someone who is very technically competent in a field like finance or engineering is given the added responsibility of managing others, for which he or she is incompetent, like trying to put a square peg in a round hole.

Some examples: during the Korean War, I was working directly for the navy under civil service rules doing varied engineering tasks, such as making sure tools used by navy scuba divers to take apart magnetic mines would not detonate them. This was in preparation for the Inchon Landing in case magnetic mines were mixed in with the contact mines. Russia had a large supply of contact mines stored close to Korea left over from their 1906 war with Japan; they gave these mines to the North Koreans. I also tested aircraft cannon, on MIG fighters the USSR had given to North Korea, for muzzle velocity, rate of fire, and projectile dispersion pattern. Dispersion, as from a shotgun, is wanted to avoid consistency of missing. I worked many days and hours for zero pay and took no vacation days. Then Congress passed a law that vacation days could not be carried over into the coming fiscal year; use them or lose them. So, still during the war, I took 30 days of vacation, driving to California and back, touring national parks. I drove my wife and her mother, to whom I owed a great debt for, in 1942, leading my wife as a teenager out of German occupied Holland, and out of Europe. They got to see America.

While Lyndon Johnson was president and a federal budget for the next fiscal year was to be established in a few months, Congress passed rules; the budget for each federal office was to be ten percent less than the budget for the present fiscal year. Government offices reacted. Offices that had been frugal went on a binge. They let contracts to have windows washed once a week, etc. Anything to avoid unspent money so they would not be cut by more than ten percent and could avoid or minimize layoffs.

An example we in the United States have experienced: a legally brilliant, but not wise lawyer, working for a Chicago civil rights law firm, sued the bank Citigroup over their refusal to issue mortgages in predominately minority populated areas where in the mortgage firm's judgment real estate prices would decline, as it had already done in similar situations in other areas. He sued on the basis that this was discriminatory against minorities who did not have the assets/income to make the mortgage payments. Most of the homes were foreclosed and many of the home buyers protested that there should be a government regulation against issuing mortgages to people who could not afford them. This civil rights case set a nationwide precedent for banks which then issued so-called subprime mortgages. Many of these were issued based on "liar loans" in which bank employees often guided mortgage applicants in lying so the banks could, on paper, comply with federal regulations. This civil rights lawsuit, which the lawyer won, greatly increased the vulnerability of banks including Citigroup and other financial institutions, plus stock markets worldwide.

The civil rights lawyer later went into politics and was elected. His name is Barrack Obama. His winning that civil rights lawsuit boomeranged against him in 2008, just as he was about to take office as president, with the financial collapse he helped create.

An example from my own experience and observation: I worked for Norden Systems, which before it became a part of United Aircraft, had designed and built the Norden bombsights of World War II. Most procurements by the US Department of Defense were cost plus fixed fee because for most military systems it was very uncertain what it would actually cost to design and then build the military systems specified. The parent company board of directors decided that company sales needed to be increased and bonuses were to be given to subsidiary presidents and heads of departments like sales and engineering as new business was brought in. Our subsidiary was bidding on two different, but similar, radars to be installed in aircraft as well as other equipment.

There were to be penalties for late deliveries. The fixed price bid on each of the two radars assumed that the design and development cost would be mostly paid for by the other contract. Then we won both contracts, were late, and lost our shirts—even eliminating most of the profit of the parent corporation. A few years later, recognizing their inability to supervise the project with a corporate vice president and a subsidiary president they had installed (but who largely created the mess) they sold the subsidiary and most of the real estate it was on.

United Technologies is regarded as a highly successful company. It was formerly named United Aircraft but changed its name after having acquired, by merger, other companies that produced air conditioners, heating systems, and elevators, thus had nothing to do with aircraft. A second reason at the time was to boost the stock price; technologies then sounded more glamorous.

It was successful in spite of gross mismanagement. I know this company from having worked there the last 35 years of my paid employment. As business was collapsing, I retired in 1990 at age 65 rather than be laid off and collect unemployment insurance and have to prove I was looking for work to collect it. My income that year went up because I collected severance pay.

Four years before I left the company I received a brass plaque mounted on wood and a small bonus for a "special award" for performance. A little earlier, the Norden vice president for sales and the vice president for engineering jointly decided to send me on a trip in order to hold down costs, ordered by corporate, to help a sister operation.

CHAPTER 13

∞

Predictions

There were many predictions that have not and will not likely come to pass. As an example, after World War II magazines such as Popular Mechanics were writing about how traffic jams would be a thing of the past. Flying cars would take people from their home garage, flying over traffic. Licensing and insurance for flying vehicles might be prohibitively expensive due to the damage they could cause.

Between World War I and World War II a suggestion that I had read in a magazine would replace young soldiers with older men. Tanks had become mechanized so youthful strength was no longer required. Young lives would be spared.

In the 1940s the thought was that there would not be a market for computers and maybe only a few would be sold.

The Wright brothers had been asked about Leonardo Da Vinci's concept of helicopters. They viewed it as not having a practical use.

A Wall Street analyst thought there was not a future market for telephones—only limited use as in intra-office communications.

Today there are predictions about weather cycles. Terminology changes: "global warming" has become "climate change." There is a belief that it is possible for humans to avoid its drastic consequences. There is a general lack of awareness that by upsetting the return flow of the Gulf Stream an early Ice Age, even possibly within the life span of many now living, may be brought on. If it is full-fledged, then glaciers could extend closer to the equator and there is even a risk of a return to "snowball earth."

Scientists are expected to save the world. Our energy problems are expected to be solved.

Although many medical challenges have been met, I disagree with the predictions of an imminent cure for cancer, as I view cancer as a variety of different problems and not as one disease. High level research using bacteria from the sea is presently being done at the Weizmann Institute. It would be wonderful to predict that this is one of the correct approaches to curing some cancers.

Although we still do not understand the means of propulsion used by UFOs, the concept of wormholes being used to travel faster than the speed of light is impossible. The aliens most likely are able to come from our galaxy by other means.

Sometimes simple inventions change our daily life for the better. Two that people found most useful and surprising were the sewing machine and the zipper. I had witnessed the coming of fiber optics. I would not have predicted it. Modern computers, the internet, the ability to travel vast distances in aircraft and the use of antibiotics were inventions which I have admired. Although I spent much of my life surrounded by inventors, I am fascinated at the concept of cameras and watches having been incorporated into cell phones. What a great idea.

AFTERWORD

Not Quite Random Thoughts, Insights & Needles from Haystacks has brought together concepts that may appear not to be connected. The thread that ties them is a sense of wonder at witnessing the universe around us. The universe is stranger than humans can imagine. Strange occurrences have happened and continue to happen. We are all part of a world larger than ourselves. It may be comfortable to simply move through life without much examination. Be engaged. Open your eyes and wonder. What are you actually witnessing? Are the interpretations of others valid? What do you think? How do you interpret the world that surrounds you? This is your puzzle to solve.

Stop and think for yourself.

ABOUT THE AUTHOR

Howard S. Halpern, CCABW, is a retired applied physicist, engineer, sometimes chemist, whose career on behalf of the United States' national defense spanned 43 years.

He worked on delivery of nuclear weapons by both missiles and aircraft, and defenses against them, including the United States' first generation intercontinental ballistic missiles (ICBMs). He was employed at United Technologies for the last 35 years of his career, retiring in 1990.

Halpern shares his life, wisdom, theories, and passions with readers.

The proud father of three daughters, grandfather, and great grandfather, Halpern lives in New England with his wife of 67 years.

www.ingramcontent.com/pod-product-compliance
Lightning Source LLC
Chambersburg PA
CBHW071429180526
45170CB00001B/279